T0213595

SpringerBriefs in Bioengineering

SpringerBriefs present concise summaries of cutting-edge research and practical applications across a wide spectrum of fields. Featuring compact volumes of 50 to 125 pages, the series covers a range of content from professional to academic. Typical topics might include: A timely report of state-of-the art analytical techniques, a bridge between new research results, as published in journal articles, and a contextual literature review, a snapshot of a hot or emerging topic, an in-depth case study, a presentation of core concepts that students must understand in order to make independent contributions.

More information about this series at https://link.springer.com/bookseries/10280

Lutz E. Claes

Mechanobiology of Fracture Healing

From Basic Science to Clinical Application

 Springer

Lutz E. Claes
Inst. f. Unfallchirurgische Forschung
und Biomechanik
Universität Ulm
Ulm, Baden-Württemberg, Germany

ISSN 2193-097X   ISSN 2193-0988   (electronic)
SpringerBriefs in Bioengineering
ISBN 978-3-030-94081-2   ISBN 978-3-030-94082-9   (eBook)
https://doi.org/10.1007/978-3-030-94082-9

This Springer imprint is published by the registered company Springer Nature Switzerland AG
The registered company address is: Gewerbestrasse 11, 6330 Cham, Switzerland

To Gaby, Sven and Paul

Preface

This book describes how biomechanical conditions in fracture fixation affect the bone healing process. The objective is to provide an overview of the state of the art in the biomechanics of fracture fixation, the mechanobiology of fracture healing, numerical simulations of fracture healing processes and specific research methods in experimental fracture healing studies as well as of the conclusions that can be drawn for clinical applications from this scientific knowledge.

This book describes the state of the art of the mechanobiology of fracture healing and is a synopsis of the scientific work of the research group of the Institute of Orthopedic Research and Biomechanics at the University of Ulm over a period of more than 40 years. The intension is to provide a source of knowledge for bio-engineers, biologists and experimental and clinical surgeons active in the field of fracture healing.

Ulm, Germany   Lutz E. Claes

Acknowledgements

I sincerely thank and acknowledge the contribution of former coworkers and PhD students. These individuals in particular include Peter Augat, Georg Duda, Christa Heigele, Anita Ignatius, Nicholaus Meyers, Sandra Shefelbine, Uli Simon, Malte Steiner, Tim Wehner, Bettina Willie and Steffen Wolf.

Contents

1   **Introduction**   1
    1.1   Introduction   1
    References   4

2   **Basic Biology of Fracture Healing**   7
    2.1   Basic Biology of Fracture Healing   7
          2.1.1   Diaphyseal Fracture Healing   8
          2.1.2   Metaphyseal Fracture Healing   10
    References   12

3   **Specific Methods in Fracture Healing Studies**   15
    3.1   Specific Methods in Experimental Fracture Healing Studies   15
          3.1.1   Animal Models for Fracture Healing Studies   15
          3.1.2   Loading Models   20
          3.1.3   Measurement of Fracture Healing Outcome   24
    References   30

4   **Basic Biomechanical Factors Affecting Fracture Healing**   35
    4.1   Basic Biomechanical Factors Affecting Fracture Healing   35
          4.1.1   Biomechanics of Fracture Fixation   35
          4.1.2   Loading of the Stabilized Bone   45
          4.1.3   Interfragmentary Movement   48
          4.1.4   Fracture Gap Size and Geometry   57
    References   59

5   **Biomechanical Enhancement of Fracture Healing**   65
    5.1   Biomechanical Enhancement of Fracture Healing   65
          5.1.1   Externally Applied Methods   65
          5.1.2   Dynamization of Fractures   70
    References   77

**6   Mechanobiological Hypotheses, Numerical Models and Their
      Application to the Improvement of Clinical Fracture Treatment**   81
      6.1   Mechanobiological Hypotheses of Tissue Differentiation
            in Fracture Healing   81
      6.2   Numerical Mechanobiological Models for Bone Healing   84
      6.3   Conclusion for the Improvement of Clinical Fracture Healing   89
      References   92

**Correction to: Basic Biomechanical Factors Affecting Fracture
Healing** C1

About the Author

Lutz E. Claes was professor of experimental surgery and biomechanics at the University of Ulm, Germany, from 1988 to 2009 and the director of the Institute of Orthopedic Research and Biomechanics from 1995 to 2009. His main research topic was the mechanobiology of fracture healing, together with other topics like biomaterials and knee biomechanics. He has published more than 500 scientific articles in peer-reviewed journals and edited several books.

Abbreviations

3D	Three-dimensional
AO/ASIF	Association for the Study of Internal Fixation
BW	Body weight
CT	Computer tomography
ESWT	Extracorporeal shockwave therapy
F	Force
GRF	Ground reaction force
IFM	Interfragmentary movement
IFS	Interfragmentary tissue strain
IS	Interlocking screw
Kfix	Stiffness of fixation
LIPUS	Low-intensity pulsed ultrasound
LMHFV	Low-magnitude high-frequency vibration
MC	Marrow cavity
pQCT	Peripheral computed tomography
R	Rotation
RM	Rotational movement
SM	Shear movement
TM	Translational movement

Chapter 1
Introduction

Abstract Fractures are one of the most frequent injuries occurring in humans. Most fractures are treated conservatively with casts and braces, but a considerable number require operative treatment. However, there remains a distinct proportion of fractures displaying complications, including delayed healing and non-unions. Complications occurring during fracture treatment can have two main reasons, the severity of the injury and the quality of the fracture treatment.

There has been considerable effort during recent decades to develop new techniques leading to advanced, minimally invasive operative techniques. With these techniques, the intraoperative trauma should be minimized to maintain the biological healing capacity as high as possible. However, with these techniques, the stability of the fracture fixation was frequently insufficient, and despite good biological conditions, the healing outcome was often unsatisfactory. To grasp the importance of biomechanics and to optimize the mechanical stability of fracture fixation, it is essential to understand the effect of fixation stiffness on the biological healing process—the mechanobiology of fracture healing. The interrelationships between the stiffness of fixation, interfragmentary movement, tissue strain, cellular reactions, tissue development and bone healing are described in this book. Clinically applied techniques are analyzed and proposals for improvements made.

1.1 Introduction

Fractures are one of the most frequent injuries occurring in every second human before the age of 65 and often require clinical treatment [1]. External splinting of fractures was the treatment of choice for centuries but with some dissatisfactory results. Currently, most fractures are treated conservatively [2] or operatively [3] by new methods of stabilization to relieve pain, prevent displacement and achieve initial stability to allow for an undisturbed bone healing. Although a great improvement of fracture treatment has been achieved in recent decades there remains a distinct proportion of fractures displaying complications including delayed healing, and non-union necessitating reoperation. Five to ten percent of the millions of fractures which occur globally develop into delayed union or non-unions [4]. This is not only an individual burden for the patients suffering from these fractures but leads to tremendous expenses for the social economic system because of the loss of working time and costs for medical treatment.

Complications occurring during fracture treatment can have two main reasons, being the severity of the injury and the quality of the fracture treatment. Accident prevention or safety precautions can reduce the severity of high-energy fractures, for example, with safer cars. However, accidents and trauma can never be completely avoided and, therefore, the quality of medical treatment and the fracture healing time have been topics of orthopedic research for a considerable time.

Unfortunately clinical diagnostic tools (i.e. X-ray) do not allow the definition of a clear healing time point as a measure of treatment success and the evaluation of the healing progress is very difficult [5]. This impedes the use of clinical studies to investigate the effect of various factors on fracture healing outcome. A better-defined clinical predictor of the quality of a medical treatment of fractures is, by contrast the rate of reoperation after a failed first fracture treatment [6, 7].

In a large study on tibia fractures, various predictors of reoperation following operative management of fractures were studied [6]. Of 20 possible prognostic variables (injury type, degree of open injury, injury mechanism, alcohol, smoking, diabetes, age, gender, pharmaceutical treatments and others) only three predictors displayed a significant correlation with the reoperation rate. These were the presence of an open fracture, remaining fracture gaps and the presence of a transverse fracture [6]. It is evident that open fractures disrupt the blood supply of the fracture healing zone and increase the risk for delayed union or non-union. Consequently this disturbs all the biological processes that require a good blood supply. It is well known from experimental studies that large fracture gaps significantly delay the healing process or even can lead to non-unions [8]. At the first sight it is not evident why simple transverse fractures so frequently require reoperation. By contrast, from a biomechanical point of view however it is clear that the stabilization of transverse fractures is often associated with a higher risk for large sheer movements than oblique or spiral fractures [9]. This is an indirect indication for insufficient biomechanical fracture stabilization as a cause for reoperation, although a measurement of the fracture stabilization was not performed in this study. However instability of fracture fixation leads to a delay of fracture healing or even non-unions, as experimentally shown in a large number of studies [10]. In particular the combination of large remaining fracture gaps and instability has long be known to be a classical non-union model in experimental studies [11]. This led us conclude that in addition to a good operative technique, three main factors influence the outcome of the fracture treatment, being the stability of fracture fixation, the fracture gap size and the biological capacity of bone healing as represented by a good blood supply [7].

There has been considerable effort during recent decades to develop new techniques and implants in the operative treatment of fractures, leading to advanced minimal invasive techniques [12]. With these techniques intraoperative trauma should be minimized and the blood supply should be preserved to maintain the biological healing capacity as high as possible. However with these techniques the stability of the fracture fixation was frequently insufficient and despite good biological conditions the healing outcome was often insufficient [13, 14].

To optimize the mechanical stability of fracture fixation it is important to understand the influence of stability on the biological healing process—the mechanobiology of fracture healing. Already in the last century it was discovered that with absolute stable conditions a direct bony union (primary healing) could occur, whereas when there was some interfragmentary movement (IFM) a stimulus for callus formation, bony bridging of the fragments (secondary healing) and a final fracture healing was observed [15]. However, in both types of fracture healing processes, unfavourable biomechanical conditions could occur which disturb the healing process or even lead to non-unions. To reduce these complications, a large number of experimental fracture healing studies have been performed in recent decades to better understand the effect of local mechanical conditions in the fracture healing zone and the corresponding cellular reactions and tissue differentiations.

From a biomechanical point of view the aim of a fracture fixation has to be, that the stability of fixation, characterized by the IFM, has to be reduced to an uncritical value, which allows the biological healing process to occur. The IFM is a function of the stiffness (stability) of fracture fixation, which can be influenced by the method of fixation and the type of implant chosen by the surgeon and the muscle activity and load bearing performed by the patient.

Unfortunately, the IFM is normally not known under clinical conditions and is not reported in clinical studies and terms like stability, rigidity and flexibility are not quantitative descriptions of the biomechanical status of the fracture fixation. Therefore, a correlation to clinical outcome parameters can hardly be used to optimize the stability of fracture fixation. The optimal scientific way to investigate the effect of the biomechanical condition of a fracture fixation on the bone healing process (mechanobiology) is, therefore, well-designed animal experiments [16, 17].

Figure 1.1 describes the influence of the stiffness of fracture fixation, the loading of the operated bone and the fracture gap size on the interfragmentary tissue strain (IFS) that occurs in the fracture gap. In addition to the biological healing capacity due to revascularization of the injured fracture zone the tissue strain level guides the tissue differentiation from hematoma to granulation tissue, bone formation and final bone healing.

In the following chapters the knowledge about these basic factors involved in the fracture healing process will be described and discussed.

The aim of this book is to provide scientists active in the field of fracture healing with the relevant information from experimental and clinical research about all the mechanobiological aspects important for the healing process. This information is necessary to analyze the meaningfulness of existing studies and to design new successful studies. In addition, it aims to provide the orthopedic surgeon with the scientific background of fracture healing and suggestions for the improvement of the clinical treatment of fractures.

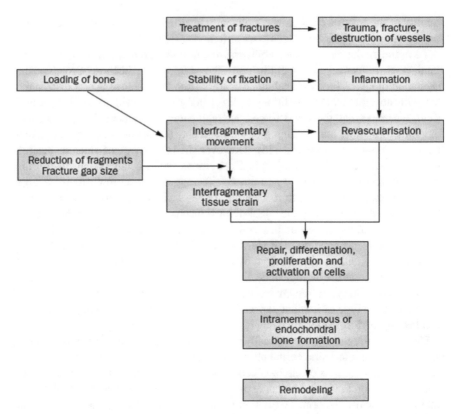

Fig. 1.1 Main factors affecting the fracture healing process. Trauma and fracture lead to blood vessel rupture inside the bone and surrounding soft tissue, cause a hematoma and initiate an inflammatory reaction. Revascularization starts and provides the healing area with cells, cytokines and growth factors. The fixation technique affects the interfragmentary movement that occurs upon loading of the bone. (IFM) causes interfragmentary tissue strain (IFS), which depends on the fracture gap size and has direct effects on the mechano-sensitive cells as well as on inflammation and revascularisation. Together these effects drive the differentiation, proliferation and activation of cells and lead to intramembranous or endochondral bone formation and healing. Finally, remodeling processes lead to reshaping of the fracture and a reconstruction of the bone (reprinted by permission from Springer Nature) [17]

References

1. Brinker, M.R., and D.P. O'Connor. 2004. The incidence of fractures and dislocations referred for orthopaedic services in a capitated population. *The Journal of Bone and Joint Surgery. American Volume* 86 (2): 290–297.
2. Sarmiento, A., and L. Latta. 1981. *Closed functional treatment of fractures*. New York: Springer Verlag.
3. Rüedi, T.P., and W.M. Murphy. 2000. *AO principles of fracture management*. Stuttgart: Georg Thieme Verlag.
4. Einhorn, T.A., and L.C. Gerstenfeld. 2015. Fracture healing: Mechanisms and interventions. *Nature Reviews Rheumatology* 11 (1): 45–54.

5. Claes, L.E., and J.L. Cunningham. 2009. Monitoring the mechanical properties of healing bone. *Clinical Orthopaedics and Related Research* 467 (8): 1964–1971.
6. Bhandari, M., et al. 2003. Predictors of reoperation following operative management of fractures of the tibial shaft. *Journal of Orthopaedic Trauma* 17 (5): 353–361.
7. Claes, L. 2021. Improvement of clinical fracture healing - What can be learned from mechano-biological research? *Journal of Biomechanics* 115: 110148.
8. Claes, L., et al. 1997. Influence of size and stability of the osteotomy gap on the success of fracture healing. *Journal of Orthopaedic Research* 15 (4): 577–584.
9. Claes, L. 2006. Biologie und Biomechanik der Osteosynthese und Frakturheilung. *Orthopädie und Unfallchirurgie* 1 (4): 329–346.
10. Claes, L.E., et al. 1998. Effects of mechanical factors on the fracture healing process. *Clinical Orthopaedics and Related Research* 355: 132–147.
11. Müller, J., R. Schenk, and H. Willenegger. 1968. Experimentelle Untersuchungen über die Entstehung reaktiver Pseudarthrosen am Hunderadius. *Helvetica Chirurgica Acta* 35: 301–308.
12. Gerber, C., J.W. Mast, and R. Ganz. 1990. Biological internal fixation of fractures. *Archives of Orthopaedic and Trauma Surgery* 109 (6): 295–303.
13. Larsen, L.B., et al. 2004. Should insertion of intramedullary nails for tibial fractures be with or without reaming? A prospective, randomized study with 3.8 years' follow-up. *Journal of Orthopaedic Trauma* 18 (3): 144–149.
14. Study to Prospectively Evaluate Reamed Intramedullary Nails in Patients with Tibial Fractures. 2008. Randomized trial of reamed and unreamed intramedullary nailing of tibial shaft fractures. *The Journal of Bone and Joint Surgery. American Volume* 90 (12): 2567–2578.
15. Willenegger, H., S.M. Perren, and R. Schenk. 1971. Primäre und sekundäre Knochenheilung. *Chirurg* 42 (6): 241–252.
16. Epari, D.R., G.N. Duda, and M.S. Thompson. 2010. Mechanobiology of bone healing and regeneration: in vivo models. *Proceedings of the Institution of Mechanical Engineers* 224 (12): 1543–1553.
17. Claes, L., S. Recknagel, and A. Ignatius. 2012. Fracture healing under healthy and inflammatory conditions. *Nature Reviews Rheumatology* 8 (3): 133–143.

Chapter 2
Basic Biology of Fracture Healing

Abstract Fracture leads to blood vessel rupture inside the bone and in the surrounding soft tissue as well as damage to cells and tissue, which together promotes the initiation of the inflammatory cascade. The acute inflammatory response occurs over the first days after fracture and is thought to initiate the repair cascade by stimulating angiogenesis, attracting and promoting the differentiation of mesenchymal stem cells and enhancing extracellular matrix synthesis. The hematoma forms a blood clot that differentiates into a granulation tissue. The most important cell types involved in bone healing (mesenchymal stem cells, osteoblasts, osteoclasts, fibroblasts and chondrocytes) create tissues like mesenchymal tissue, bone, cartilage and connective tissue.

In diaphyseal fractures, the repair phase begins with intramembranous bone formation at the periosteum some distance from the fracture, which drives callus formation. The callus expanse and progresses in diameter and direction to the fracture, which it finally bridges and stabilizes.

Cancellous bone in metaphyseal areas appears to heal with no or very limited callus formation but rather by a direct bone formation on existing trabeculae. Following bony bridging, the bone remodeling phase occurs that finally leads to a normal bone structure.

2.1 Basic Biology of Fracture Healing

The biology of bone healing involves a highly complex system of a large number of molecules and genes [1, 2]. A considerable proportion of these factors are affected by the mechanobiological environment in the fracture healing zone [3, 4]. Ode et al. found more than a hundred genes are up-regulated or down-regulated when a fracture fixation in a rat osteotomy model was compared between a rigid and a semi-rigid fixation [4].

However, very little is known how these genes effect cellular differentiation and cell activity as a function of the biomechanical environment as well as possible interactions between the various genes in regard to bone healing.

As an output of all these highly complex reactions, by contrast the histological images of the fracture healing zone show the cell and tissue differentiation, extracellular matrix production and vascularization. Therefore, the final tissue formation in the healing zone can be analyzed as a function of the tissue strain without knowing the complex genetic regulative processes.

Depending on the amount of tissue strain applied to a fracture gap, the most important cell types involved in bone healing (mesenchymal stem cells, osteoblasts, osteoclasts, fibroblasts and chondrocytes) and tissues like mesenchymal tissue, bone, cartilage and connective tissue can be analyzed [5]. In this chapter the mechanobiology of various fracture types as a function of the mechanical environment will be described on the bases of histological images and vascularization.

2.1.1   Diaphyseal Fracture Healing

Fracture healing in diaphyseal bones under flexible fixation follows a characteristic course, which can be divided in to three partially overlapping phases: inflammation, repair and remodeling [2] (Fig. 2.1).

Fracture leads to blood vessel rupture within the bone and in the surrounding soft tissue as well as damage to cells and tissue, which promotes the initiation of the inflammatory cascade [1, 2, 6]. The acute inflammatory response occurs over the first days after fracture and is thought to initiate the repair cascade by stimulating angiogenesis, attracting and promoting the differentiation of mesenchymal stem cells and enhancing extracellular matrix synthesis. The hematoma forms a blood clot that differentiates into a granulation tissue. The repair phase begins with intramembranous bone formation at the periosteum some distance from the fracture,

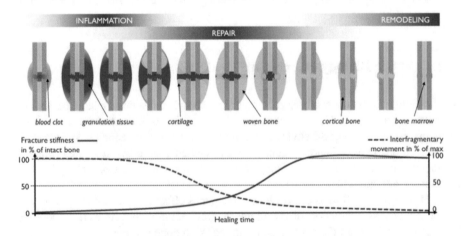

Fig. 2.1 Phases of fracture healing, morphological development of callus formation, course of fracture stiffness and interfragmentary movement (IFM) versus healing time

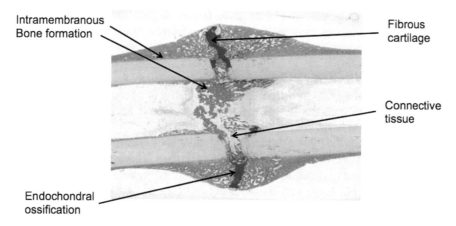

Intramembranous
Bone formation

Fibrous
cartilage

Connective
tissue

Endochondral
ossification

Fig. 2.2 Longitudinal section through a callus healing in a sheep tibia fracture (Paragon staining, magnification 4×) (reprinted by permission from John Wiley and Sons) [7]

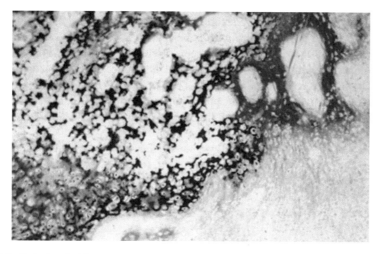

Fig. 2.3 Endochondral bone healing at the front of callus formation in flexible fixed fractures. Paragon staining, magnification 100× (reprinted by permission from Elsevier) [8]

which drives callus formation (Figs. 2.1 and 2.2). The callus grows and progresses in diameter and direction towards the fracture.

At a larger distance from undamaged vessels hypoxic conditions and larger tissue strain stimulate only chondrocyte proliferation, the formation of fibrous cartilage and endochondral ossification (Fig. 2.3) to woven bone [2, 8]. Following peripheral bony bridging of the fracture, the callus diameter decreases and bone is remodeled. Basic multicellular units resorb the woven bone and build a lamellar bone structure. Osteoclasts resorb the callus and reshape the bone to its physiological size.

The stiffness of the healing area increases with time from very low values in the inflammatory phase followed by a progressive increase in stiffness during the repair phase up to values in the remodeling phase that are comparable to or even higher than the stiffness of the corresponding uninjured bone (Fig. 2.1). The interfragmentary movement (IFM) displays a contrary course than the stiffness course of the healing bone. Immediately after fracture fixation the IFM is maximal and mainly influenced by the stiffness of the fixation device. When the repair phase starts and the callus develops from granulation tissue to fibrocartilage and subsequently to woven bone the IFM is steadily reduced and is finally very small after bony bridging and in the remodeling phase (Fig. 2.1).

It is well established that a moderate IFM stimulates callus formation in diaphyseal fractures, which can lead to a rapid healing process [2]. Very large IFMs due to unstable fixation or overloading of the bone may stimulate the formation of a large callus formation but avoid a rapid bony bridging of the fracture line [9–12] and can thus cause a hypertrophic non-union. Very small IFMs can suppress the stimuli for bone formation and can thus delay the healing process [13–16].

The amount by which healing can be stimulated is dependent on the amplitude and duration of the IFM applied. A large IFM in the early, inflammatory phase of healing does not significantly affect the bony volume of callus tissue, but induces more cartilage formation in the fracture callus [10, 12, 17, 18] and might delay bone healing. The main increase in callus volume appears to occur in the repair phase of healing [12, 19, 20] and the amount thereof is dependent on the IFM [21, 22]. Following bony bridging of the callus the IFM is minimal and the callus volume decreases through remodeling processes [12, 19, 20]. The best amplitude of the IFM and the optimal time for its application to stimulate the most rapid fracture healing however, is dependent on the individual fracture situation and will be discussed in the following chapters.

When the IFM is extremely low and the fragments are under contact and compression or the fracture gap is less than 1 mm [23], a direct, primary bone healing can occurs [24]. Osteons cross the fracture and connect both fragments (Fig. 2.4). This direct bone healing occurs without any external callus formation, bypasses the reparative bone healing phase and starts directly with the remodeling phase (Fig. 2.5).

2.1.2 Metaphyseal Fracture Healing

Although many clinical fractures occur in metaphyseal areas of the bone, only a limited number of experimental studies have been performed to characterize the mechanobiology of bone healing in the metaphyseal area [26–28]. It is generally supposed that metaphyseal fractures heal differently compared with fractures in the bone diaphysis. In diaphyseal fractures the cortical bone is injured and the dominant type of fracture healing is callus healing. Cancellous bone in metaphyseal areas appear to heal with no or very limited callus formation [29, 30]. However, there appears to be a greater healing capacity in the metaphyseal bone. Possible reasons for the better healing in the metaphyseal region might be its larger active bone

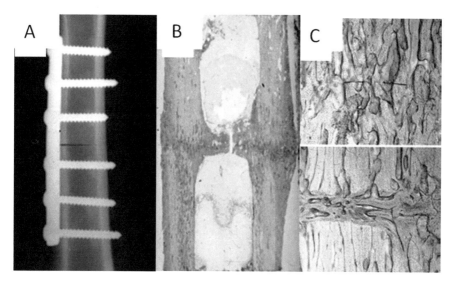

Fig. 2.4 Primary diaphyseal fracture healing in a sheep metatarsel osteotomy model under extremely low IFM. (**a**) Radiograph of the osteotomy after rigid fixation using a compression plate. Bone healing occurs without any external callus formation. (**b**) Longitudinal histological section 24 weeks after surgery (Paragon staining, magnification 4×). Gap healing (adjacent to the plate, left) or contact healing (opposite to the plate, right) between the fragments. (**c** top) Higher magnification of the cortical contact healing area. The fracture line is directly crossed by osteons (magnification 20×, Paragon staining). (right below) A gap of limited size is first filled by woven bone, subsequently bridged by osteons growing in the longitudinal direction (Permission obtained from Lippincott Wiliams & Wilkins, Philadelphia 2005) [25]

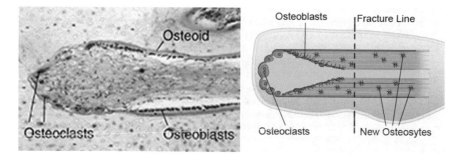

Fig. 2.5 Bone remodeling units (osteons) that can cross and connect bone fragments under close contact and mechanically very stable fixation (minimal IFM). Osteoclasts resorb a tunnel across the fracture line that is filled with newly formed bone (osteoid) by osteoblasts in several layers (lamellas) (left: histological section, right: sketch) (reprinted by permission from Georg Thieme Verlag) [20]

surface, the better blood supply, and its thicker periosteum containing more cells [31]. The fracture healing occurs by bone apposition on existing trabeculae that grow in the direction of the fracture gap or fracture line (Fig. 2.6). This can occur by

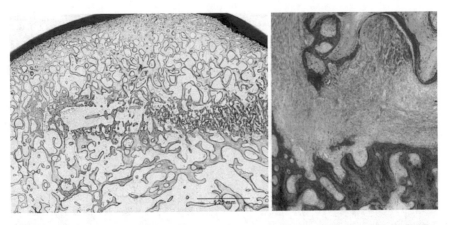

Fig. 2.6 Metaphyseal bone healing in the condyle of the sheep femur. Left trabecular bone healing closes a 3mm fracture gap without an internal or peripheral callus formation [27] (reprint by permission from Georg Thieme Verlag). Right: higher magnification of trabecular bone healing in a fracture gap with active osteoblast seams (light blue) on existing trabeculae (reprinted by permission from Springer Nature) [26]

intramembranous bone formation or by endochondral bone healing depending on the mechanical environment in the fracture zone.

References

1. Einhorn, T.A. 1998. The cell and molecular biology of fracture healing. *Clinical Orthopaedics* 355: 7–21.
2. Claes, L., S. Recknagel, and A. Ignatius. 2012. Fracture healing under healthy and inflammatory conditions. *Nature Reviews Rheumatology* 8 (3): 133–143.
3. Palomares, K.T., et al. 2009. Mechanical stimulation alters tissue differentiation and molecular expression during bone healing. *Journal of Orthopaedic Research* 27 (9): 1123–1132.
4. Ode, A., et al. 2014. Interaction of age and mechanical stability on bone defect healing: an early transcriptional analysis of fracture hematoma in rat. *PLoS One* 9 (9): e106462.
5. Cullinane, D.M., et al. 2002. Induction of a neoarthrosis by precisely controlled motion in an experimental mid-femoral defect. *Journal of Orthopaedic Research* 20 (3): 579–586.
6. McKibbin, B. 1978. The biology of fracture healing in long bones. *Journal of Bone and Joint Surgery* 60-B(2): 150–162.
7. Claes, L., et al. 2008. Temporary distraction and compression of a diaphyseal osteotomy accelerates bone healing. *Journal of Orthopaedic Research* 26 (6): 772–777.
8. Claes, L.E., and C.A. Heigele. 1999. Magnitudes of local stress and strain along bony surfaces predict the course and type of fracture healing. *Journal of Biomechanics* 32 (3): 255–266.
9. Augat, P., et al. 2003. Shear movement at the fracture site delays healing in a diaphyseal fracture model. *Journal of Orthopaedic Research* 21 (6): 1011–1017.
10. ———.1996. Early, full weightbearing with flexible fixation delays fracture healing. *Clinical Orthopaedics* 328: 194–202.
11. Claes, L., et al. 1997. Influence of size and stability of the osteotomy gap on the success of fracture healing. *Journal of Orthopaedic Research* 15 (4): 577–584.

12. Epari, D., et al. 2006. Instability prolongs the chondral phase during bone healing in sheep. *Bone* 38 (6): 864–870.
13. Bottlang, M., et al. 2010. Far cortical locking can improve healing of fractures stabilized with locking plates. *The Journal of Bone and Joint Surgery. American Volume* 92 (7): 1652–1660.
14. Claes, L. 2011. Biomechanical principles and mechanobiologic aspects of flexible and locked plating. *Journal of Orthopaedic Trauma* 25 (Suppl 1): S4–S7.
15. Röderer, G., et al. 2014. Delayed bone healing following high tibial osteotomy related to increased implant stiffness in locked plating. *Injury* 45 (10): 1648–1652.
16. Hente, R., et al. 1999. Fracture healing of the sheep tibia treated using a unilateral external fixator. Comparison of static and dynamic fixation. *Injury* 30 (Suppl 1): A44–A51.
17. Lienau, J., et al. 2005. Initial vascularization and tissue differentiation are influenced by fixation stability. *Journal of Orthopaedic Research* 23 (3): 639–645.
18. Miclau, T., et al. 2007. Effects of delayed stabilization on fracture healing. *Journal of Orthopaedic Research* 25 (12): 1552–1558.
19. Wu, J.J., et al. 1984. Comparison of osteotomy healing under external fixation devices with different stiffness characteristics. *The Journal of Bone and Joint Surgery. American Volume* 66 (8): 1258–1264.
20. Claes, L. 2006. Biologie und Biomechanik der Osteosynthese und Frakturheilung. *Orthopädie und Unfallchirurgie* 1 (4): 329–346.
21. Claes, L.E., et al. 1995. Effect of dynamization on gap healing of diaphyseal fractures under external fixation. *Clinical biomechanics* 10 (5): 227–234.
22. ———. 1998. Effects of mechanical factors on the fracture healing process. *Clinical Orthopaedics and Related Research* 355: S132–S147.
23. Marsell, R., and T.A. Einhorn. 2011. The biology of fracture healing. *Injury* 42 (6): 551–555.
24. Perren, S.M. 1979. Physical and biological aspects of fracture-healing with special reference to internal-fixation. *Clinical Orthopaedics and Related Research* 138: 175–196.
25. Claes, L.E., and K. Ito. 2005. Biomechanics of fracture fixation and fracture healing. In *Basic orthopaedic biomechanics and mechano-biology*, 3rd ed. V.C. Mow and R. Huiskes, 563–584. Philadephia, PA: Lippincott Williams & Wilkins.
26. Claes, L., et al. 2009. A novel model to study metaphyseal bone healing under defined biomechanical conditions. *Archives of Orthopaedic and Trauma Surgery* 129 (7): 923–928.
27. ———. 2011. Modelle der metaphysären Frakturheilung. *Osteologie* 20 (1): 29–33.
28. ———. 2011. Metaphyseal fracture healing follows similar biomechanical rules as diaphyseal healing. *Journal of Orthopaedic Research* 29 (3): 425–432.
29. Jarry, L., and H.K. Uhthoff. 1971. Differences in healing of metaphyseal and diaphyseal fractures. *Canadian Journal of Surgery* 14 (2): 127–135.
30. Uhthoff, H.K., and B.A. Rahn. 1981. Healing patterns of metaphyseal fractures. *Clinical Orthopaedics* 160: 295–303.
31. Fan, W., R. Crawford, and Y. Xiao. 2008. Structural and cellular differences between metaphyseal and diaphyseal periosteum in different aged rats. *Bone* 42 (1): 81–89.

Chapter 3
Specific Methods in Fracture Healing Studies

Abstract A knowledge of some specific methods is important when fracture healing experiments are planned, performed and analyzed. In addition to general guidelines for experimental research, there are specific biomechanical and histomorphological methods.

Besides large animal models (sheep, dogs and rabbits) that have been used successfully for a considerable time, the use of small animal models (rats and mice) has become more popular because of lower costs and the possibility to investigate the molecular processes during bone healing. The different physiological loading and bone healing of various animals is described and advantages and restrictions of these models are discussed. The creation of a standardized fracture model is limited to small animals under specific conditions, whereas for large animals, the standardized osteotomy is the method of choice. Loading of the fractured bone model can be by unrestricted or restricted walking or by externally applied forces that create a certain interfragmentary movement.

Methods to apply and measure forces and interfragmentary movements are described. Various methods to measure the fracture healing outcome like micro-computed tomography stiffness and strength of healed bones, histomorphological methods to determine the various tissues in the fracture healing zone and methods to determine the structure and amount of blood vessels are described.

3.1 Specific Methods in Experimental Fracture Healing Studies

3.1.1 Animal Models for Fracture Healing Studies

In addition to the general requirements related to experimental animal models [1], there are some specific demands that have to be taken into consideration when fracture healing studies are planned that attempt to investigate the mechanobiology of fracture healing. One of the most important choices is the selection of the animal species.

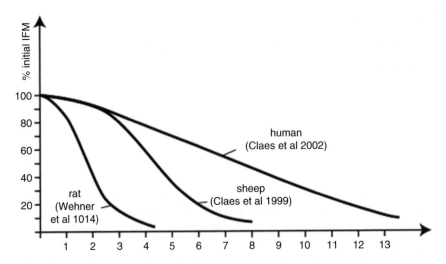

Fig. 3.1 Characteristic examples of the course of IFM versus healing time for various species under flexible fixation and callus healing of diaphyseal fractures. The decreasing IFM indicates the progress of the bone healing process

Previously a considerable number of different animal models have been used to study the mechanobiology of fracture healing [2]. In particular large animals like sheep [3–7] and dogs [8, 9] but also rabbits were used [2, 10–12]. Larger animal models are preferred because their bone healing is more similar to that of humans [2, 13], and the application of surgical techniques, adaptation of clinically used stabilization systems and the biomechanical testing of the healed bones is easier in larger animals. The use of dogs depends on the societal acceptance of their role in research and have been less frequently used in the last years whereas sheep are increasingly used.

In recent decades the use of rodents like rats and mice has become more important [14]. Rats are easy to handle and cost much less than large animals. This allows studies with larger numbers of animals. In addition, it is possible to apply more molecular biological methods.

Mice became more popular because of availability of a broad spectrum of antibodies and gene-targeted animals that allowed the study of molecular mechanisms of fracture healing [14, 15]. However, to studying the mechanobiology of fracture healing, the standardization and control of the biomechanics of fracture fixation is very important and this is very frequently a challenge with rodent fracture healing models [14].

Although the character of the course of IFM versus the healing time in diaphyseal callus healing is similar for all species used for experimental studies, the time scale is very different (Fig. 3.1). The healing is more rapid for small animals than for large animals and humans (Fig. 3.1). But even for the same species the time course of interfragmentary movement (IFM) is different for animals of different strains, fracture fixations with various stiffness and animal models with different age,

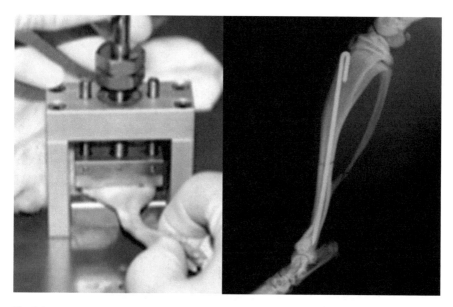

Fig. 3.2 (left) Apparatus to break the tibia of the rat (reprinted by permission from Springer Nature) [16]. (right) Transverse fracture of the rat tibia after Kirschner' wire fixation (reprinted by permission from Acta Orthopaedica) [17]

fracture type, level of injury, specific site, fracture gap size, soft tissue damage, local vascularity and gender [2].

Moreover, the time course of the IFM versus the healing time is available only for a few experimental studies. Therefore, a comparison of the results of various experimental studies has to be made very critically and conclusions drawn carefully.

The creation of a standardized fracture in an animal bone is a challenge and not possible for all species and bones. For rats and mice it is possible to drop a weight with a defined energy to the bone and create a closed fracture of the tibia [18] that preserves the hematoma [19]. In our experience in the majority of the cases (approximately 90%) a simple transverse fracture will occur in rat and mouse tibia fracture models (Fig. 3.2). The remaining fractures display an oblique fracture or several bony fragments. However, this method does not successfully work on the femur with a larger muscle cover of the bone. Therefore, most of the studies use an open approach to the femur and an osteotomy of the bone.

A comparative study in rabbit tibiae [20] between a closed fracture and an osteotomy model showed that the callus formation and early strength and stiffness recovery after two weeks was better in the fracture group, but after four weeks there was no longer a significant difference.

In larger animals like sheep, an osteotomy is the dominating technique to create a "standardized fracture model" [6]. The advantage of an osteotomy model is the standardization of the fracture geometry but the disadvantage is the greater tissue damage of the muscles, periosteum and intramedullary vessels.

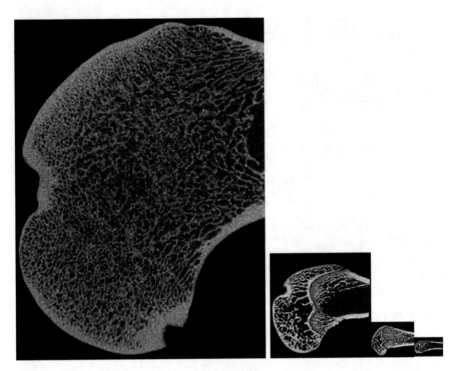

Fig. 3.3 Micro-computed tomography images of femoral condyles of sheep, rabbit, rat and mouse (left to right) (magnification 2×) demonstrating the tremendous differences in size (adapted from [28]) (reprinted by permission from Georg Thieme Verlag)

To achieve an experimental fracture in a certain area of a sheep bone it is necessary to create a predetermined breaking point by a partial osteotomy or a drill hole to create stress riser [21–27]. A partial osteotomy of a sheep tibia and subsequent manual artificial fracturing however led to a random fracture type [22]. Transverse, oblique or even wedge-shaped fractures occur [22]. In regard to the fracture healing the authors stated that healing progress for an oblique diaphyseal osteotomy might be comparable to bone healing of a random fracture if the intactness of the periosteum is ensured.

The diversity of fracture types may, however, cause higher standard deviation in healing outcome parameters than osteotomies and could make it more difficult to find statistical differences between experimental groups.

Experimental studies on metaphyseal fracture models were performed on sheep, dogs, rabbits and rats mainly in the femoral condyle but also in the proximal and distal tibia. Published studies on metaphyseal models were very frequently limited to radiological and histological analyses without a characterization of the biomechanical conditions in the healing zone [28].

It is nearly impossible to establish a metaphyseal fracture model in rodents because of the small size of the condyles (Fig. 3.3). Most of the trabecular bone

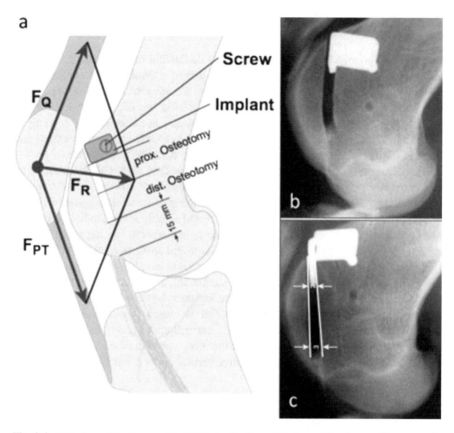

Fig. 3.4 Metaphyseal fracture model. (**a**) Under loading of the knee joint the quadriceps muscle force FQ and the counteracting patella force FPT result in a patello-femoral contact force FR, which deflects the trochlea and causes a movement in the osteotomy. The movement is limited by the implant in the proximal gap. (**b**) X-ray of the sheep knee joint under unloaded conditions. A 3 mm wide and parallel osteotomy is provided. (**c**) X-ray of the sheep knee with a 2 mm implant under weight-bearing of the leg. The trochlea is deflected till it attaches against the implant. The 3 mm osteotomy gap (distal) is approximately linear and one constricted to 2 mm at the contact to the implant (reprinted by permission from John Wiley and Sons) [29]

can be found in the condyles but the number of trabeculae there is very limited. In addition, the growth plate is located in the condyle and this should not be disturbed by a fracture or osteotomy or by implant materials. Otherwise, an unphysiological bone formation and growth occurs that overlaps the normal bone healing process.

Recently, a first sheep model with characterized biomechanical conditions could demonstrate a correlation between the interfragmentary strain (IFS) and bone healing similar to diaphyseal fractures (Fig. 3.4) [29].

3.1.2 Loading Models

There are in principle two main possibilities of loading the experimental fracture model. Individually loading the animal leg due to ground reaction forces. This method corresponds more to the normal clinical situation with human patients and can lead to clinically relevant results. The disadvantage of this method for the analyses of the healing process is that it is normally not known what loading or IFM occurs at which time period of the healing process at the fracture level and that animals display a wide range of individual loading behavior (like humans).

The exact determination of the course of loads or IFM versus healing time as an important biomechanical factor for the healing process is difficult. Monitoring of IFM directly at the fracture [30] or indirectly at the fracture fixation system [31, 32] is technically demanding and has been performed in only a few studies (Fig. 3.5). The measurement of the ground reaction force underneath the leg using special measuring platforms [33–35] is possible for large and small animals [36] (Fig. 3.8). It is difficult to estimate the load at the fracture level based on the ground reaction force [37, 38] (Figs. 3.6, 3.7, and 3.8). For the loading at the fracture level the ground reaction force and the muscle forces of the leg must have to be taken into consideration [38, 39]. The ground reaction force and the muscle forces change depending on the gait phase and the muscle forces are normally not known. In general, it is a question of which load should be chosen that best represents the loading during the entire gait cycle. Normally, the maximum vertical ground reaction force is used,

Fig. 3.5 Indirect measurement of IFM during gait at the metatarsel osteotomy gap of the sheep hind limb. The ring fixator enables the stabilization of the metatarsel osteotomy against bending and torsion but while allowing adjustable axial movements (left) by a telescoping system connecting the proximal and distal rings (right). The fixator is instrumented by a displacement sensor that allows the measurement of the axial IFM during loading [32]. A telemetry system on the back of the sheep (left), transmits the data to the laboratory

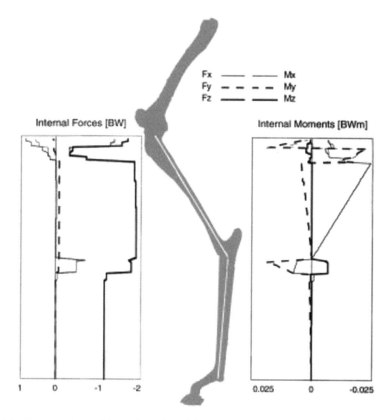

Fig. 3.6 Maximum internal forces in body weight (BW) and moments in BWm during the stance phase of sheep gait (maximal ground reaction force) computed with muscle forces obtained from linear (dark lines) and non-linear optimization (light lines). Z-axis: longitudinal axis of bone, X-axis in the anterior direction, Y-axis in the medial direction (reprinted by permission from Elsevier) [39]

even forces in other directions and moments around the three main axes can affect the IFM resulting from such complex loading [37, 38]. The determination of the muscle forces is only possible by numerical musculo-skeletal models [38, 39].

As indicated above, the loading of the experimental fracture is individually very different and can lead to unwanted overloading of the fracture fixation, that is, when an animal is jumping (what normally does not occur with a patient). To avoid such uncontrolled loading events and standardize the loading sometimes the animals are restricted in loading theire legs. They can be suspended in slings to compensate the main body weight (BW) and allow only a partial weight bearing (like in patients) [40]. Another possibility is to cut the Achilles tendon, which prevents an immediate postoperative loading of the lower extremity but allows a slow and continues increase in loading with the spontaneous healing of the tendon [33] (like for increasing partial weight bearing often recommended for patients). The second

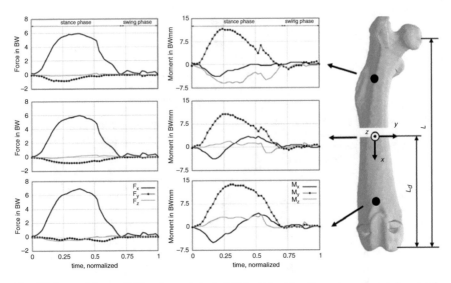

Fig. 3.7 Internal forces (left) and moments (middle) at three locations along the rat femur axis (right) during the entire whole gait cycle (reprinted by permission from Elsevier) [38]

Fig. 3.8 Custom-made gait wheel with pressure sensors (top) to measure the ground reaction force during gait (bottom left); custom-made external fixator equipped with strain gauge to measure the deformation at the fixation device and indirectly the IFM in the fracture gap during gait (bottom right) (reprinted by permission from John Wiley and Sons) [36]

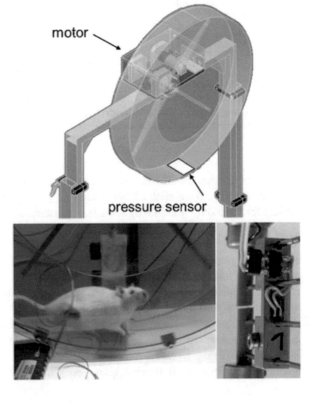

Fig. 3.9 Stabilization of a sheep tibia osteotomy by a unilateral external fixator. A motor-driven external stimulation unit was temporarely attached to the inner fixator bars and coused a cyclic opening of the osteotomy gap due to bending of the fixator [41]

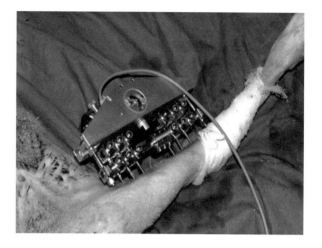

method is unrelated to clinical conditions, but does allows the study of the influence of enforced and constant IFM on the healing process. This enables the control of the displacement, number of load cycles and frequency of cyclic loading. This is possible with externally applied movements by mechanically driven loading systems [7, 41–44] (Fig. 3.9). This can be interesting when the effect of constant IFM to the tissue differentiation in bone healing or the effect of the number of load cycles should be studied. In this model the forces acting in the bone healing zone will increase with increasing tissue formation and stiffness of the growing tissue. It has to be taken into consideration that sometimes an overlapping of displacement controlled and physiological loading can occur when the animals are allowed to load their legs after externally applied displacement has ended [7, 41]. When the fracture is stabilized under contact of the fracture surfaces and an application of an IFM is not possible a force-controlled application of loads to a fracture healing model [45, 46] is possible. Therefore, the type of model that will be used for a fracture healing study has to be selected depending on the specific research question.

3.1.3   Measurement of Fracture Healing Outcome

Unfortunately, there is no exact definition for when a fracture is healed. Under clinical conditions the assessment of fracture healing is performed on the basis of radiographs where the callus healing in long bone fractures is visible in terms of callus bridging and the absence of the fracture line. However, the assessment of fracture healing is, subjective and does not allow reliable determination of bone healing. Studies have shown that radiographic analyses are subjective and inaccurate, that the amount of callus does not correlate to the stiffness of the healed bone [47], and cortical bridging poorly predict the bone healing results [48].

For experimental studies there are in principle four main methods to characterize the fracture healing outcome, new quantitative radiological images, like computer tomography (CT), to show the calcified bone formation, biomechanical measurements of stiffness and strength, histomorphological images to characterize the tissue and cell composition and measurements of revascularization.

3.1.3.1   Radiological Images

To characterize the bone healing by X-rays is the classical method in clinical settings and is frequently used in experimental studies, but can only provide an overview of the healing status. The amount of new callus formation and the number of bridged cortices can only estimate the progress in bone healing and is not applicable for metaphyseal fractures. A better resolution and quantification of fracture healing images is possible by the quantitative peripheral computed tomography (pQCT) technique or micro-CT, often available today. In contrast to X-ray images, these methods allow the analyses of the three-dimensional (3D) distribution and quantification of the calcified bone formation in the healing zone [49, 50]. A difficulty of this method is the definition at which gray level the visible tissue is considered to be calcified bone and thus it normally requires the use of a standard with a known hydroxyapatite density for calibration [49]. The volumetric bone mineral density, determined by such methods, at the level of the fracture zone strongly correlates with biomechanical findings [49]. Higher magnifications are possible with modern micro-CT systems and allow images in different plains (Fig. 3.10).

The combination of 3D density measurements with voxel based finite element analysis to assess the fracture stiffness in healing calluses is a sophisticated method to determine the stiffness of bone healing (Fig. 3.11) without sacrificing the animals [50].

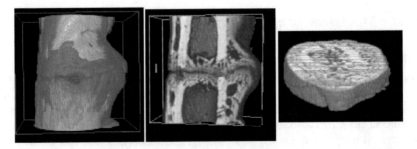

Fig. 3.10 Micro-CT images in three-dimensional presentation (left), longitudinal (middle) and cross-section (right) of a fracture healing specimen from a rat femur

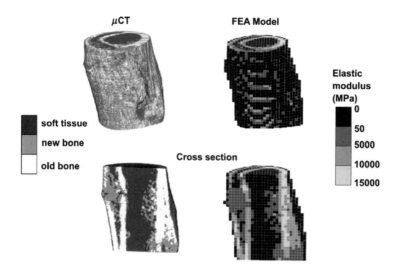

Fig. 3.11 Creating a voxel finite element model from micro-CT scans. The elastic modulus was linearly scaled to the gray level value of the bone region (reprinted by permission from Elsevier) [50]

The use of micro magnetic resonance imaging (MRI), that is very seldom available, can be very beneficial when in addition to the calcified tissue the cartilage and connective tissue should be analyzed [51].

3.1.3.2   Biomechanical Measurements

The measurement of stiffness and strength of the healing bone is an important factor because it allows the clinically relevant estimation of how loadable the bone is at a certain healing phase.

Stiffness and strength correlate quite well in the early fracture healing phase [52]. In the late healing phase (bone remodeling) the stiffness might even exceed that of the normal healthy bone, because of the large diameter of the callus, but the normal strength is not achieved [52].

Normally, callus formation around a diaphyseal fracture is not symmetrical. This causes some difficulties in regard to mechanical testing. Formulae for the calculation of strength and stiffness assume known loading axis, cross sectional area and center of gravity for loaded cross sections. This information is usually not known for the individually different fracture healing zones.

The anatomically irregular form of bones makes it difficult to fix them in the testing machine in a standardized manner leading to angular and axial deviations from the "assumed" bone axis.

Different test methods, including torsion and bending tests, are well established. Each method offers advantages and disadvantages, and it is still debated as to which of these is the most sensitive to experimental conditions like specimen alignment, directional dependency and asymmetric behavior. To clarify these questions, various test specimens were investigated with different test methods to determine the sensitivity of the tests [53]. Five standardized test specimens with different geometries, mimicking size, directional behavior and embedding variations of respective rat long bone specimens were tested. The mechanical tests included three-point bending, four-point bending, cantilever bending, axial compression (Fig. 3.12) and constrained and unconstrained torsion. The bending tests were highly dependent on the rotational direction of the maximum fracture callus expansion relative to the loading direction whereas angular deviations were negligible [53]. Compared to four-point bending, three-point bending is easier to apply on small rat and mouse bones under realistic testing conditions and yields robust measurements, provided there is low variation of the callus shape among the tested specimens. Axial compression testing was highly sensitive to embedding variations, and therefore cannot be recommended. Although unconstrained torsion testing is experimentally difficult to realize, it was found to be the most robust method, because it was

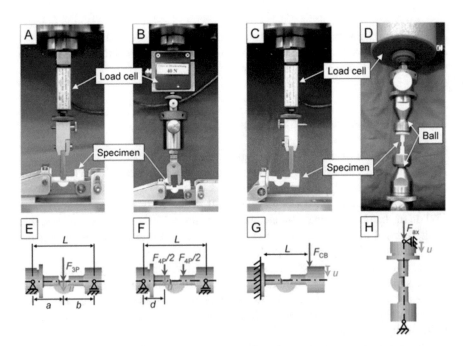

Fig. 3.12 Test setups and boundary conditions for four different mechanical test methods for rat tibia models with asymmetrical callus. (**a**) and (**e**) three-point bending, (**b**) and (**f**) four-point bending, (**c**) and (**g**) cantilever bending, (**d**) and (**h**) axial compression. L, a, b, d, lever arms for applied forces F, u: deflection of the specimen (reprinted by permission from PlosOne) [53]

independent of both rotational alignment and embedding uncertainties. Constrained torsion testing showed only small errors because of a parallel offset between the specimens´ axis of gravity and the torsional axis of rotation [53].

The determination of the bending stiffness requires the measurement of the deflection of the bone under load. This should be performed directly at the bone that is, with optical measuring methods. The displacement recorded by the testing machine includes other deformations, that is, in the setting of the experimental construction and the indentation of the loading device in the callus surface, and could lead to incorrect stiffness data.

3.1.3.3 Histomorphological Images

Following alcohol fixation and dehydration bone specimens can be embedded undecalcified in plastic or after decalcification in paraffin. From the bone blocks histological sections can be prepared. In undecalcified sections, the calcified bone is clearly visible but the thickness of the sections is normally relatively great (30–70 µm). This limits the judging of cell morphology but does allow, with suitable surface staining, a very good discrimination of the various tissues and cell types involved in fracture healing (Fig. 3.13). Paraffin histology allows the production of very thin sections for a large number of staining methods including immuno-histological

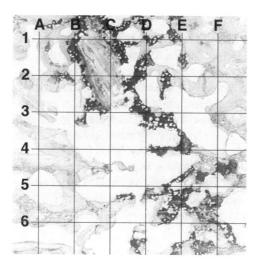

Fig. 3.13 Point counting method for quantifying histomorphological images. Using a grid ocular, the area of interest was divided by vertical and horizontal lines. The tissue visible underneath the crossing points of the grid were analyzed and counted. The horizontal line 4, for example, shows the following tissues at the vertical lines: intramembranous bone formation (**a, b**); connective tissue (**c, d**); endochondral ossification (**e**); and connective tissue (**f**) (reprinted by permission from Wolters Kluwer Health) [54]. Undecalceilied bone slice (70 µm) with Paragon staining (Toluidin blue and Fuchsin)

Fig. 3.14 Longitudinal
section through the
peripheral callus of a sheep
metatarsal after osteotomy
and flexible fixation
(9 weeks p.o.). Fluorescence
light microscopy,
demonstrating two phases of
callus formation, (green:
calcein green (4 weeks),
yellow: reverin (8 weeks)).
Artificial red lines indicating
the beginning of the 4-week
and 8-weeks bone
formation, demonstrate the
growth in callus formation.
Adapted from [58]
(reprinted by permission
from Elsevier)

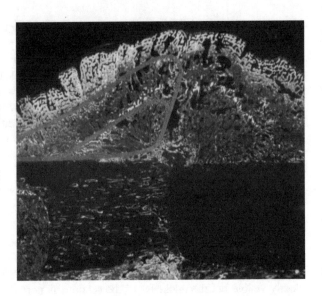

staining, and better cellular analyses. The quantification of the histological images is
elaborate. Normally, the two-dimensional histological sections are quantified by
software supported image analyzing systems or by manually performed point
counting methods (Fig. 3.13) [54, 55]. A large number of points have to be taken
into consideration in the later method to achieve a statically representative result for
the areas covered with the various tissues [55].

A very elaborate method is the 3D characterization of the bone healing area using
a large number of serial consecutive sections from the entire fracture healing
zone [56].

The fluorescence labeling of bone formation allows the determination of bone
growth at various time points during the bone healing process [57]. Following the
injection of fluorescent dyes, these are incorporated in the newly formed calcified
bone approximately 2–4 days after injection. Under fluorescent light microscopy
undecalcified sections of bones can be analyzed (Fig. 3.14). There are several dyes
available that create different colors, but ultraviolet lights and filters with different
wavelengths need to be used in the microscope [57]. Not all dyes are clearly visible
and when using a single filter, a restricted selection of only two to three dyes is
meaningful. This practically limits the application to a small number of dyes.

3.1.3.4   Vascularization

Blood flow can be measured by laser Doppler flowmetry in vivo at different tissue
depths [59]. However, the penetration of the laser light is limited. In soft tissue, the
blood flow can be measured at a depth of up to 6mm, but in bone only up to 2 mm.
This limits the application of this method to small animal studies on rats and mice. It

Fig. 3.15 The rat lower hind limb with a tibia fracture, during measurement of the blood flow using a laser Doppler probe with standardized contact forces to the skin (reprinted by permission from Springer Nature)

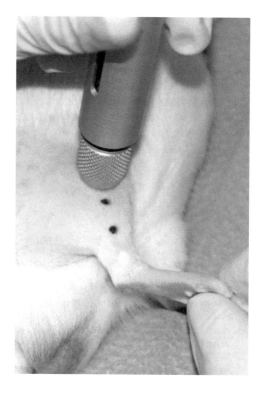

has to be taken into consideration that the application force of the laser probe to the skin (Fig. 3.15) can influence the skin vascularization. For our applications a custom-made measuring head with a spring inside the holder was developed to ensure a standardized application force of the laser probe [59]. The great advantage of this method is that non-invasive longitudinal studies with different measuring time points during the bone healing process can be performed.

Blood vessels in bone can be shown when a contrast medium like barium sulfate is injected in the main artery of the extremity [60] and the bone specimen decalcified (microangiography). High resolution X-rays made from slices of the bone specimen can demonstrate the vessel architecture in one plane of the bone healing area. Currently, micro-CT techniques can be applied to determine the 3D architecture and vessel distribution in a bone healing area of interest. Instead of barium sulfate India ink can be used to show vessels in standard histological sections [61]. Because artifacts like incomplete filling of vessels can occur and the quantification of the vessels is difficult with both methods, these methods are seldom employed at present.

Blood vessels without contrast medium can be detected in standard thin histological section with an adequate staining and quantitatively determined with histomorphological methods [62, 63]. The question is however how representative the selected section of the bone is in regard to the blood supply of the entire bone.

The total blood flow in a bone volume can be measured by injecting radioactive microspheres in bone penetrating vessels (Scintigraphy). The radioactivity in a bone specimen can be measured using a scintillation spectrometer and is representative of the blood flow in the volume of interest [64, 65].

Whereas laser Doppler flowmetry and scintigraphy determine the blood flow in a tissue volume, the microangiography and histomorphology of vessels allow the analyses of the local vessel distribution and vessel size in one plane.

References

1. Auer, J.A., et al. 2007. Refining animal models in fracture research: seeking consensus in optimising both animal welfare and scientific validity for appropriate biomedical use. *BMC Musculoskeletal Disorders* 8: 72.
2. O'Loughlin, P.F., et al. 2008. Selection and development of preclinical models in fracture-healing research. *The Journal of Bone and Joint Surgery. American Volume* 90 (Suppl 1): 79–84.
3. Nunamaker, D.M., and S.M. Perren. 1979. A radiological and histological analysis of fracture healing using prebending of compression plates. *Clinical Orthopaedics and Related Research* 138: 167–174.
4. Perren, S.M., et al. 1973. Developments of compression plate techniques for internal fixation of fractures. *Progress in Surgery* 12: 152–179.
5. Claes, L., et al. 1997. Influence of size and stability of the osteotomy gap on the success of fracture healing. *Journal of Orthopaedic Research* 15 (4): 577–584.
6. Epari, D.R., G.N. Duda, and M.S. Thompson. 2010. Mechanobiology of bone healing and regeneration: in vivo models. *Proceedings of the Institution of Mechanical Engineers* 224 (12): 1543–1553.
7. Goodship, A.E., and J. Kenwright. 1985. The influence of induced micromovement upon the healing of experimental tibial fractures. *Journal of Bone and Joint Surgery* 67B (4): 650–655.
8. Olerud, S., and G. Danckwardt-Lilliestrom. 1968. Fracture healing in compression osteosynthesis in the dog. *Journal of Bone and Joint Surgery. British Volume (London)* 50 (4): 844–851.
9. Chao, E.Y., et al. 1989. The effect of rigidity on fracture healing in external fixation. *Clinical Orthopaedics* 241: 24–35.
10. Veneroni, G., B. Boccadoro, and F. Pluchino. 1962. Fixation of P-32 in the focus of a fracture and in osseous callus in the long bones in rabbits. *Archivio di Ortopedia* 75: 1338–1341.
11. Lettin, A.W. 1965. The effects of axial compression on the healing of experimental fractures of the rabbit tibia. *Proceedings of the Royal Society of Medicine* 58 (11): 882–886.
12. Park, S.H., et al. 1998. The influence of active shear or compressive motion on fracture-healing. *Journal of Bone and Joint Surgery* 80 (6): 868–878.
13. Wissing, H., and M. Stürmer. 1986. Untersuchungen zur Knochenregeneration nach Unterbrechung der medullären oder periostalen Strombahn bei verschiedenen Versuchstier-Species. *Hefte zur Unfallheilkunde* 181: 225–228.
14. Histing, T., et al. 2011. Small animal bone healing models: standards, tips, and pitfalls results of a consensus meeting. *Bone* 49 (4): 591–599.
15. Garcia, P., et al. 2013. Rodent animal models of delayed bone healing and non-union formation: a comprehensive review. *European Cells & Materials* 26: 12.

16. Claes, L., et al. 2017. The effect of a combined thoracic and soft-tissue trauma on blood flow and tissue formation in fracture healing in rats. *Archives of Orthopaedic and Trauma Surgery* 137 (7): 945–952.

17. ———. 2011. The effect of both a thoracic trauma and a soft-tissue trauma on fracture healing in a rat model. *Acta Orthopaedica* 82 (2): 223–227.

18. Bonnarens, F., and T.A. Einhorn. 1984. Production of a standard closed fracture in laboratory animal bone. *Journal of Orthopaedic Research* 2 (1): 97–101.

19. Nunamaker, D.M. 1998. Experimental models of fracture repair. *Clinical Orthopaedics and Related Research* 355: S56–S65.

20. Park, S.H., et al. 1994. Technique for producing controlled closed fractures in a rabbit model. *Journal of Orthopaedic Research* 12 (5): 732–736.

21. Tepic, S., et al. 1997. Strength recovery in fractured sheep tibia treated with a plate or an internal fixator: an experimental study with a two-year follow-up. *Journal of Orthopaedic Trauma* 11 (1): 14–23.

22. Dumont, C., et al. 2009. Long-term effects of saw osteotomy versus random fracturing on bone healing and remodeling in a sheep tibia model. *Journal of Orthopaedic Research* 27 (5): 680–686.

23. Baumgaertel, F., S.M. Perren, and B. Rahn. 1994. Animal experiment studies of "biological" plate osteosynthesis of multi-fragment fractures of the femur. *Unfallchirurg* 97 (1): 19–27.

24. Hente, R., et al. 1999. Fracture healing of the sheep tibia treated using a unilateral external fixator. Comparison of static and dynamic fixation. *Injury* 30 (Suppl 1): A44–A51.

25. Park, S.H., et al. 1999. Comparison of healing process in open osteotomy model and closed fracture model. *Journal of Orthopaedic Trauma* 13 (2): 114–120.

26. Schemitsch, E.H., et al. 1998. Quantitative assessment of bone injury and repair after reamed and unreamed locked intramedullary nailing. *The Journal of Trauma* 45 (2): 250–255.

27. Decker, S., et al. 2014. Non-osteotomy and osteotomy large animal fracture models in orthopedic trauma research. *Orthopaedic Review* 6 (4): 5575.

28. Claes, L., et al. 2011. Modelle der metaphysären Frakturheilung. *Osteologie* 20 (1): 29–33.

29. ———. 2011. Metaphyseal fracture healing follows similar biomechanical rules as diaphyseal healing. *Journal of Orthopaedic Research* 29 (3): 425–432.

30. Klein, P., et al. 2004. Comparison of unreamed nailing and external fixation of tibial diastases–mechanical conditions during healing and biological outcome. *Journal of Orthopaedic Research* 22 (5): 1072–1078.

31. Gardner, T.N., et al. 1994. Three-dimensional movement at externally fixated tibial fractures and osteotomies during normal patient function. *Clinical biomechanics* 9 (1): 51–59.

32. Claes, L.E., et al. 1995. Effect of dynamization on gap healing of diaphyseal fractures under external fixation. *Clinical biomechanics* 10 (5): 227–234.

33. Augat, P., et al. 1996. Early, full weightbearing with flexible fixation delays fracture healing. *Clinical Orthopaedics* 328: 194–202.

34. Recknagel, S., et al. 2011. Experimental blunt chest trauma impairs fracture healing in rats. *Journal of Orthopaedic Research* 29 (5): 734–739.

35. Röntgen, K.V., et al. 2010. Fracture healing in mice under controlled rigiid and flexible conditions. In *International society for fracture repair 12th biennial meeting*. London: International Society for Fracture Repair.

36. Wehner, T., et al. 2014. Temporal delimitation of the healing phases via monitoring of fracture callus stiffness in rats. *Journal of Orthopaedic Research* 32 (12): 1589–1595.

37. Duda, G.N., et al. 1998. A method to determine the 3-D stiffness of fracture fixation devices and its application to predict inter-fragmentary movement. *Journal of Biomechanics* 31 (3): 247–252.

38. Wehner, T., et al. 2010. Internal forces and moments in the femur of the rat during gait. *Journal of Biomechanics* 43 (13): 2473–2479.

39. Duda, G.N., et al. 1998. Analysis of inter-fragmentary movement as a function of musculoskeletal loading conditions in sheep. *Journal of Biomechanics* 31 (3): 201–210.

40. Bishop, N.E., et al. 2006. Shear does not necessarily inhibit bone healing. *Clinical Orthopaedics and Related Research* 443: 307–314.
41. Augat, P., et al. 2001. Mechanical stimulation by external application of cyclic tensile strains does not effectively enhance bone healing. *Journal of Orthopaedic Trauma* 15 (1): 54–60.
42. Hente, R., et al. 2004. The influence of cyclic compression and distraction on the healing of experimental tibial fractures. *Journal of Orthopaedic Research* 22 (4): 709–715.
43. Cullinane, D.M., et al. 2002. Induction of a neoarthrosis by precisely controlled motion in an experimental mid-femoral defect. *Journal of Orthopaedic Research* 20 (3): 579–586.
44. Tufekci, P., et al. 2018. Early mechanical stimulation only permits timely bone healing in sheep. *Journal of Orthopaedic Research* 36 (6): 1790–1796.
45. Gardner, M.J., et al. 2008. Pause insertions during cyclic in vivo loading affect bone healing. *Clinical Orthopaedics and Related Research* 466 (5): 1232–1238.
46. Takeda, T., T. Narita, and H. Ito. 2004. Experimental study on the effect of mechanical stimulation on the early stage of fracture healing. *Journal of Nippon Medical School* 71 (4): 252–262.
47. Sano, H., et al. 1999. Correlation of radiographic measurements with biomechanical test results. *Clinical Orthopaedics and Related Research* 368: 271–278.
48. McClelland, D., et al. 2007. Fracture healing assessment comparing stiffness measurements using radiographs. *Clinical Orthopaedics and Related Research* 457: 214–219.
49. Augat, P., et al. 1997. Quantitative assessment of experimental fracture repair by peripheral computed tomography. *Calcified Tissue International* 60 (2): 194–199.
50. Shefelbine, S.J., et al. 2005. Prediction of fracture callus mechanical properties using micro-CT images and voxel-based finite element analysis. *Bone* 36 (3): 480–488.
51. Haffner-Luntzer, M., et al. 2017. Evaluation of high-resolution in vivo MRI for longitudinal analysis of endochondral fracture healing in mice. *PLoS One* 12 (3): e0174283.
52. Chehade, M.J., et al. 1997. Clinical implications of stiffness and strength changes in fracture healing. *Journal of Bone and Joint Surgery. British Volume (London)* 79 (1): 9–12.
53. Steiner, M., et al. 2015. Comparison between different methods for biomechanical assessment of ex vivo fracture callus stiffness in small animal bone healing studies. *PLoS One* 10 (3): e0119603.
54. Claes, L., A. Ruter, and E. Mayr. 2005. Low-intensity ultrasound enhances maturation of callus after segmental transport. *Clinical Orthopaedics and Related Research* 430: 189–194.
55. Weibel, E.R. 1963. Principles and methods for the morphometric study of the lung and other organs. *Laboratory Investigation* 12: 131–155.
56. Gerstenfeld, L.C., et al. 2006. Three-dimensional reconstruction of fracture callus morphogenesis. *The Journal of Histochemistry and Cytochemistry* 54 (11): 1215–1228.
57. Rahn, B.A. 1976. Die polychrome sequenzmarkierung des knochens. *Nova Acta Leopoldina* 44: 249–255.
58. Claes, L.E., and C.A. Heigele. 1999. Magnitudes of local stress and strain along bony surfaces predict the course and type of fracture healing. *Journal of Biomechanics* 32 (3): 255–266.
59. Bumann, M., et al. 2003. Influence of haemorrhagic shock on fracture healing. *Langenbeck's Archives of Surgery* 388 (5): 331–338.
60. Rhinelander, F.W. 1974. Tibial blood supply in relation to fracture healing. *Clinical Orthopaedics* 105: 34–81.
61. Chidgey, L., et al. 1986. Vascular reorganization and return of rigidity in fracture healing. *Journal of Orthopaedic Research* 4 (2): 173–179.

62. Claes, L., K. Eckert-Hübner, and P. Augat. 2003. The fracture gap size influences the local vascularization and tissue differentiation in callus healing. *Langenbeck's Archives of Surgery* 388 (5): 316–322.
63. Lienau, J., et al. 2005. Initial vascularization and tissue differentiation are influenced by fixation stability. *Journal of Orthopaedic Research* 23 (3): 639–645.
64. Wallace, A.L., et al. 1994. The vascular response to fracture micromovement. *Clinical Orthopaedics* 301: 281–290.
65. Grundnes, O., and O. Reikeras. 1992. Blood flow and mechanical properties of healing bone. Femoral osteotomies studied in rats. *Acta Orthopaedica Scandinavica* 63 (5): 487–491.

Chapter 4
Basic Biomechanical Factors Affecting Fracture Healing

Abstract The stabilization of a fracture is biomechanically characterized by the stiffness of the fixation/bone compound. It can be determined by the measurement of the interfragmentary movement under loading of the fractured and fixed bone. Characteristic stiffness data of typical fixation methods, like intramedullary nails, external fixators and plate constructs, are described for stabilizations in human patients and animal experiments. Data from load-bearing experiments in humans and animals are described and the resulting interfragmentary movements shown. Monitoring the interfragmentary movement and load bearing versus time as a method to quantify the bone healing course are described

The effect of interfragmentary movement on the interfragmentary tissue strain and its influence on tissue differentiation and revascularization in the bone healing area are discussed. Besides the amount of interfragmentary movement, the effect of the direction of movement and the important role of the fracture gap size for tissue differentiation in the fracture healing area are shown.

4.1 Basic Biomechanical Factors Affecting Fracture Healing

4.1.1 Biomechanics of Fracture Fixation

The biomechanics of fracture fixation describes the mechanical behavior of the fixation configuration under relevant loading conditions. It is normally characterized by the stiffness of the construct or the interfragmentary movement (IFM) under various loading directions. Most fractures are treated with plaster casts or braces (splinting). The operative treatment of fractures is, however, increasingly performed using a large number of fixation devices in trauma and orthopedic surgery. Because the biomechanics of fracture fixation, in particular, is important in the operative treatment

---

The original version of this chapter was revised: Figure 4.6 has been updated. A correction to this chapter can be found at https://doi.org/10.1007/978-3-030-94082-9_7

L. E. Claes, *Mechanobiology of Fracture Healing*, SpringerBriefs in Bioengineering, https://doi.org/10.1007/978-3-030-94082-9_4

of fractures to achieve a successful bone healing, this book is focused on the operative treatment of fractures using the most important types of implants and techniques.

The stiffness of a fixation construct cannot be determined directly. It is calculated by dividing the load applied to the fracture fixation (force or moment) by the corresponding deformation (displacement or rotational angle). The stiffness of a fracture fixation construct can generally be determined in three loading directions and around three rotation axes. A complete six-degree-of-freedom assessment of construct stiffness is available only for a small number of fixation constructs [1, 2]. Stiffness measurements are most often performed in vitro using cadaver bones or plastic bone models because in vivo measurements on patients or animals are very difficult to realize.

In vitro stiffness measurements of fracture fixations on cadaver bones or even plastic bone models are only simplified descriptions of the reality in patients. Simplified loading directions and load values in comparison to physiological conditions together with the absence of muscle forces can result in very different stiffness values compared to the reality. Even the number of the restraint loading directions in vitro tremendously influences the stiffness data of the implant. In addition, it has to be taken into consideration that the measured IFM is normally related to the center of the fracture gap and the IFM at the periphery of the bone can be different (i.e. under bending loads) and are effected by the constrains of the mechanical test [3]. In addition to this limitation, it is evident that the chosen loading characteristic can only represent one example from the great variety of loading activities that a patient or an animal can perform. An improvement of the loading conditions can be achieved when force vectors are chosen that simulate forces in bones and muscles under different activities [4]. This would allow the selection of critical loading conditions for a fixation construct to test. In addition most often the contribution of other load carrying structures, like for example the fibula at the lower leg and the soft tissue envelop of the entire leg [5] are not considered when, for example, only the isolated tibia is tested. Therefore the in vitro determination of construct stiffness only estimates the actual construct stiffness present clinically.

4.1.1.1 Characteristic Stabilities of Human Fracture Fixations

Stiffness of Intramedullary Nail Constructs

There are two biomechanically different types of intramedullary nails, the reamed and the unreamed nail. The reamed nail is a large-diameter slotted nail for press-fit insertion into a reamed marrow cavity whereas the unreamed nail is a small-diameter nail, the diameter of which is much smaller than the inner (endosteal) diameter of the bone. Translational shear movements in reamed nails can be nearly zero and the shear stiffness very high when the marrow cavity is reamed in the zone of the fracture location and the nail is inserted in a press-fit manner. Shear forces perpendicular to the long axis of the bone can cause low shear stiffness and large shear movement between the unreamed nail and bone that is as large as the difference in the nail diameter and the inner bone diameter (Fig. 4.1a). For an intramedullary canal

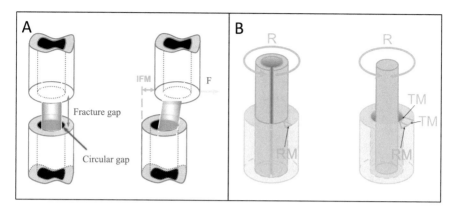

Fig. 4.1 (**a**) Sheer stiffness of unreamed intramedullary nails is low because only small forces (F) are necessary to move the nail relative to the bone (interfragmentary movement (IFM)). The IFM in the translational direction depends on the circular gap between the inner diameter of the bone and the nail diameter. (**b**) For a reamed slotted intramedullary nail (left side) fitted tight into the inner diameter of the bone the main interfragmentary movement results from the rotation (R). An unreamed intramedullary nail (right side) with a diameter smaller than the inner diameter of the bone allows a translational movement (TM) that adds to the shear movement caused by the rotation (RM) (reprinted by permission from Georg Thieme Verlag) [6]

of 12-mm diameter, a 9-mm diameter nail yielded a shear stiffness of 131 N/mm, and an 11-mm nail had a shear stiffness of 224 N/mm [7]. The axial stiffness was 723 N/mm for the 9-mm nail and 1039 N/mm for the 11-mm nail [8]. Another study reported axial stiffness in the range of 1500–2300 N/mm for 8–12 mm diameter nails. Far greater axial stiffness results of 4000 N/mm were measured for an 11-mm reamed nail [5]. The translational shear stiffness is particular critical for transverse and multifragmentary fractures, because with low shear stiffness, even small forces can shift a small-diameter unreamed nail within the larger diameter of the intramedullary canal until the nail contacts the endosteal surface (Fig. 4.1a). This effect is even more pronounced when the fracture is not located at the isthmus of the bone, rather further proximal or distal where the intramedullary canal widens.

In addition to the low shear stiffness, the small diameter unreamed nail display also a low rotational stiffness resulting in an additional shear movement under torsion of the stabilized bone (Fig. 4.1b). Under physiological torsional moments, rotational angles of 6–47° between fracture surfaces have been reported, depending on the nail type and diameter [8, 9]. The shear movements caused by translational shear forces and torsional loads are superimposed in the fracture zone and can cause critical interfragmentary tissue deformations (Fig. 4.1b). The stiffness of nail fixation is additionally influenced by motion between interlocking screws and interlocking holes in the nail (Fig. 4.2a). To reduce shear motion due to toggle of interlocking screws inside the nail, angle stable interlocking systems have been developed [9, 11].

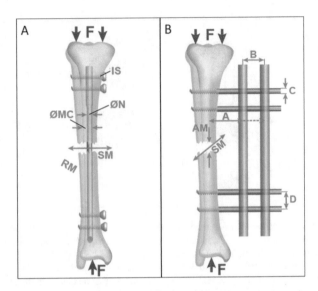

Fig. 4.2 (**a**) The critical stiffness of unreamed intramedullary nails depends mainly on the translational shear movement (SM) in the medio-lateral and anterior-posterior directions, the difference between the nail diameter (N) and diameter of the marrow cavity (MC), and the rotational movement (RM). The bending stiffness can be increased when an angle-stable interlocking screw (IS) is used [10]. (**b**) The axial movement (AM) and shear movement (SM) of the fixator under load depend mainly on the distance between the stabilization element and bone (A), the distance between the stabilization elements (B), the diameter of the bone screws (C) and the distance between the screws (D) (adapted from Claes [10]) (reprinted by permission from Springer Nature)

Stiffness of External Fixator Constructs

External fixation is primarily used for temporary stabilization of open or infected fractures. Despite its temporary application, sufficient fixation stiffness is important to initiate the bone healing process, which may under favorable conditions proceed to successful fracture healing without the need for revision to an alternate fixation type. Most external fixator systems can be applied in a considerably variety of configurations that can greatly differ in stiffness [12].

For long bone fractures unilateral external fixator systems are used most frequently. The stiffness of unilateral external fixators mainly depends on the size, configuration and material of the fixator elements and bone screws. Fixation stiffness can mainly be increased by several means (Fig. 4.2b): by decreasing the distance between the stabilization element and the bone; by increasing the distances between the stabilization elements and the bone pins; or by using larger diameter bone screws (Fig. 4.2b).

Only a few clinical measurements on the axial stiffness of unilateral fixator exist, documenting an axial stiffness of 174 N/mm [13], approximately 240 N/mm [14] and 510 N/mm [15].

In vitro investigations of the AO/ASIF-double tube system yielded axial stiffness reports of 148 N/mm [16], 250 N/mm [5] and 364 N/mm [17]. Sheer motion perpendicular to the diaphysis is usually of a similar order of magnitude than the axial motion under physiological loading.

The stiffness of ring-fixator constructs depends mainly on the number of rings, ring diameter, wire diameter and number of screws. Axial-stiffness values of 33–117 N/mm [16] and torsional-stiffness values of 0.7–1.5° have been reported [18]. Ring fixators have therefore a lower axial stiffness than unilateral fixators but a far greater torsional stiffness than unilateral external fixators and intramedullary nail constructs.

Most of these studies were performed on isolated tibiae. With an intact fibula and soft-tissue coverage, the stiffness in vivo can be assumed to be approximately one third higher [5] and is believed to improve the healing process [19].

Stiffness of Plate Constructs

Fracture fixation with conventional compression plates relies on a very high stiffness of the implant-bone construct, which can result in direct bridging of bone fragments by osteonal healing processes [20]. However, a loss of compression between the fragments can cause some IFM with large tissue strains and consequently bone resorption [20]. In recent years, the use of locking plates has become more frequent. These locked plating constructs do not compress the fragments together but retain a fracture gap and achieve stability by fixed angle screw fixation [20, 21]. Bending of the plate causes underneath the plate IFM and tissue strain, that are extremely low, leading to the suppression of the bone-formation stimulus. For very stiff locking plates, an axial stiffness of up to 3000 N/mm was measured [22]. Reducing the bending stiffness of the plating construct by increasing the bridge span [21] increases IFM and interfragmentary strain (IFS) only opposite the plate, but not appreciably beneath the plate (Fig. 4.3a, b). An offset of the plate from the bone surface could increase the tissue strain in the fracture gap both near and opposite the plate, but would also decrease the load to failure of the plate under cyclic loading [25]. Measurements of the translational shear motion beneath the plate are only known in the plane of the screws and remain small provided the bridge span is short. For a long bridge span, shear motion dominates axial motion [22]. However, shear motion can occur opposite the plate, because, of the low rotational stiffness of the plate [21]. The asymmetric axial motion at the fracture gap makes it difficult to define an axial and shear stiffness (Fig. 4.3a, b). It is therefore more practical to evaluate the interfragmentary movements individually, at specific locations of the fracture. In vitro studies of locking plate constructs showed 0.8 mm and 1 mm axial IFM at the cortex opposite the plate in response to 250 N and 350 N axial loading, respectively. However, motion beneath the plate remained minimal [23]. Assuming a patient with partial load bearing of the leg and a clinically relevant fracture gap of 2.5 mm, the interfragmentary movement opposite the plate causes a tissue strain of approximately 30%. Such an IFS is known to stimulate callus formation [23]. Should a plate fixation lead to contact between the bony fragments opposite of the plate, the entire

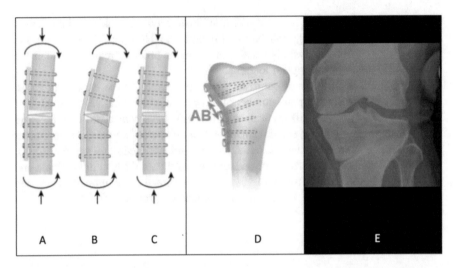

Fig. 4.3 Interlocking plate fixations in diaphyseal fractures (**a-c**) and metaphyseal osteotomies (**d, e**). Interlocking plate fixations with a remaining fracture gap lead under loading to a strain distribution (blue area) with some interfragmentary tissue strain opposite the plate but minimal tissue strain directly beneath the plate (**a**). Reducing the stiffness of plate fixation by using a smaller number of screws leads to a larger tissue strain opposite the plate, but still very low strains beneath the plate (**b**). From a mechanobiological point of view, a uniform strain distribution across the fracture gap would be good for the stimulation of bone healing in all fracture-healing zones (**c**) (reprinted by permission from Walters Kluwer Health) [23]. (**d**) An open-wedge osteotomy of the tibia head stabilized by a stiff plate leads to minimal interfragmentary movement (AB) and tissue strain beneath the plate [10] (reprinted by permission from Springer Nature), which still suppresses bone formation 18 months after an operation; visibly after plate removal (**e**) (reprinted by permission from Elsevier) [24]

construct becomes very stiff, with virtually no interfragmentary movement or tissue strain beneath the plate. For example, after stabilizing an open-wedge osteotomy of the proximal tibia with a plate applied to bridge the open wedge, a 500 N axial load induced only 0.05 mm of axial motion adjacent to the plate (Fig. 4.3d), corresponding to a stiffness of 10,000 N/mm [24]. Accordingly, a clinical study on plating of high tibia osteotomies reported that no bone healing was observed in 66% of all patients directly beneath the plate (Fig. 4.3e) [24].

4.1.1.2 Characteristic Stabilities of Experimental Fracture Fixations

The stiffness of experimental fracture fixations (normally osteotomies as standardized fracture models) is most often better defined as under clinical conditions and standardized within an experimental group. Similarly like in clinical studies, the deformation in all 6-degrees of freedom under defined loads, which allows the calculation of 3D stiffness, however, is described only for fixation systems for sheep [26] and rats [27].

Table 4.1 Stiffness of fracture fixation devices for sheep experimental models

	Axial stiffness (N/mm)	Shear stiffness (N/mm)
External fixators		
Goodship et al. [28]	500	
Goodship et al. [28]	700	
Claes et al. [29]	5000[a]	
Duda et al. [1]	425	
Wolf et al. [30]	183	170
Augat et al. [31]	850	
Klein et al. [32]	1959	
Hente et al. [33]	1666	
Bishop et al. [34]	498	
Epari et al. [35]	2540	164
Epari et al. [35]	2177	433
Epari et al. [35]	1523	347
Epari et al. [35]	1479	344
Schell et al. [36]	650	
Tufekci et al. [37]	770[a]	
Tufekci et al. [37]	10,000[a]	
Tufekci et al. [37]	20,000[a]	
Plates		
Bottlang et al. [38]	3922[a]	2500[a]
Bottlang et al. [38]	628[a]	600[a]
Intramedullary nails		
Klein et al. [32]	1775	
Epari et al. [35]	1213	139
Epari et al. [35]	2762	469

[a]Calculated from published data

Most often the axial stiffness and in some studies the shear stiffness (translational or rotational) or bending stiffness is determined. To estimate the critical IFM the axial and shear stiffness are the most relevant data (Tables 4.1, 4.2, and 4.3).

Fracture healing studies are performed in a number of different species. Large animal studies are performed mainly in sheep [52] and dogs [53]. The advantage of large animal models is that the bone healing process is more similar to that in human bone healing [54–56] than for small animal models and that often standardized implants for human application can be used. In recent decades mainly sheep models have been used (Fig. 4.4), but dog models only very seldom because mixe-bread dogs differ in regard to bone healing and inbread dogs are very expensive, together with a declining willingness of society to accept canine experiments.

Studies with rabbits are relative rare [59, 60], whereas studies with rats and mice have increased considerably in recent decades [61]. Rats are easy to handle and cheap and mice have the great advantage that there is a broad spectrum of antibodies and gene-targeted animals that allow the study of molecular mechanisms of fracture

Table 4.2 Stiffness of fracture fixation devices for rat experimental models

	Axial stiffness (N/mm)	Rotational stiffness (Nmm/o)
External fixator		
Mark et al. [39]	61	
Mark et al. [40]	30	
Mark and Ryderik [41]	265	
Strube et al. [42]		
Offset 5 mm	41	9
Offset 10 mm	19	7
Offset 15 mm	10	6
Willie et al. [43]		
Offset 6 mm	138	14
Offset 8 mm	77	12
Offset 10 mm	46	12
Recknagel et al. [44]		
Rigid, 6 mm offset	119	
Flexible, 12 mm offset	31	
Glatt et al. [45]	254	
	185	
	114	
	25	
Osagie-Clouard et al. [46]	30	47
	6	19
Harrison et al. [47]		46
Plates		
Boerkel et al. [48]	350	
	214	
Nottebaert et al. [49]		40
		52

Table 4.3 Stiffness of fracture fixation devices for mice experimental models

	Axial stiffness (N/mm)	Rotational stiffness (Nmm/o)
External fixator		
Histing et al. [50]		1.52
Röntgen et al. [51]	18.10	1.50
Röntgen et al. [51]	0.82	1.20
Locking plate		
Histing et al. [50]		1.10
Intramedullary nail		
Histing et al. [50]		
Conventional nail (pin, needle)	0	
Locking nail		0.19

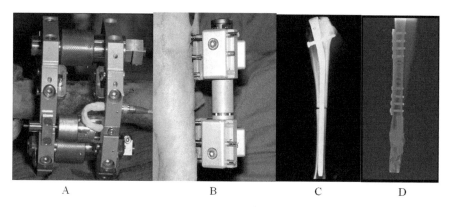

A   B   C   D

Fig. 4.4 Fracture healing models in sheep studies. (**a**) Very rigid ring fixator with adjustable axial movement and sensor for the measurement of IFM (reprinted by permission from Elsevier) [57]. (**b**) Unilateral external fixator with adjustable axial and rotational movements [31]. (**c**) Interlocking nail [58]. (**d**) Bridging plate (reprinted by permission from Springer Nature) [58]

Fig. 4.5   Rat fracture fixation designs: (**a**) UlmExFix, (**b**) RatExFix, (**c**) RatFix, (**d**) RatNail (reprinted by permission from PlosOne) [27]. The UlmExFix [62] is custom made, the remaining systems commercially available (RiSystem, Davos, Switzerland)

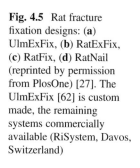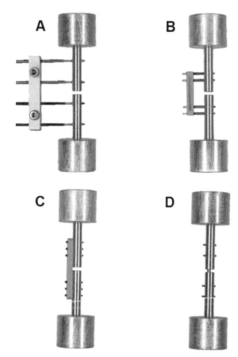

healing [61]. For a long time, the standard fixation technique for small animal fracture models was intramedullary fixation using wires or injection needles (Figs. 4.5 and 4.6) [50]. When it was realized that this form of fixation did not achieve any rotational or axial stability, which significantly disturbed the healing process [50], the development of more sophisticated implant systems was initiated.

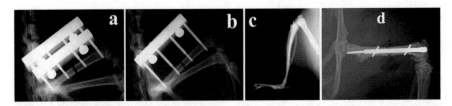

Fig. 4.6 Frequently used fixation techniques for rat fractures. For the femur most often an unilateral external fixator is used (**a** stiff fixator, **b** flexible fixator) [43, 62] (reprinted by permission from John Wiley and Sons). For the tibia frequently injection needles or steel wires are used (**c**) as an intamedullary nail (reprinted by permission from John Wiley and Sons [63]. For the femur interlocking nails are now available (**d**) (reprinted by permission from John Wiley and Sons) [64]

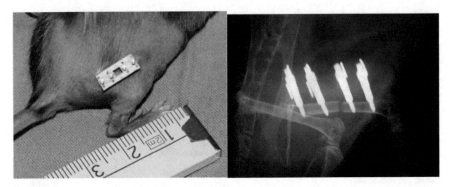

Fig. 4.7 External fixation for mice femur osteotomy. (left) Rigid polymer fixator with titanium screws at the lateral side of the femur. (right) X-ray of the fracture fixation (reprinted by permission from John Wiley and Sons) [51]

External fixators, plates and interlocking nails were subsequently developed [50] (Tables 4.2 and 4.3; Figs. 4.5, 4.6, and 4.7).

For all animal fracture healing models a large number of different fixation devices have been used. Most of the studies did not characterize the stiffness of their fracture fixation method. Tables 4.1, 4.2, and 4.3 describes the stiffness data that are published or could be calculated from existing data. These data show the great variation in axial and shear stiffness between various studies and explain why most often the results of various studies can hardly be compared.

Table 4.1 demonstrates the tremendous variation of the axial stiffness for external fixators used for fracture healing studies in sheep ranging over 183–20,000 N/mm. The translational shear stiffness is generally low, but not so greatly deviating (164–344 N/mm). The stiffness of plates is high, ranging in the axial direction over 628–3922 N/mm and in the shear direction over 600–2500 N/mm. The nails are very stiff under axial loads (1775–2762 N/mm) whereas they have very low translational shear stiffness (139–469 N/mm). In addition to the translational stiffness the rotational stiffness (not reported) increases the combined shear movement under the load resulting from translation and rotation (see Fig. 4.1b).

For the rat fracture fixations (Table 4.2) the axial stiffness of external fixators shows, like for the sheep models, an extremely large variation over 6–265 N/mm. The rotational shear stiffness ranges over 6–47 N mm/degree. Plates display a relatively high axial stiffness.

For mice there are only a few stiffness data published (Table 4.3). External fixators rang over 0.82–18.1 N/mm axial stiffness and 1.2–1.52 Nmm/degree rotational shear stiffness. The locking plate displays a similar rotational stiffness as external fixators (1.1 Nmm/degree) whereas the commonly used technique for intramedullary pin fixation by wires or needles do not exhibit any rotational stiffness.

Only the new developed interlocking nail showed some rotational stiffness that was however much smaller than for external fixators (0.19 Nmm/degree) [50].

4.1.2   Loading of the Stabilized Bone

4.1.2.1   Forces and Moments at the Normal Leg

A direct measurement of the load acting at the fracture site in vivo is not possible. Instrumented fracture fixation devices can be used to measure the load at the device immediately post operatively when fracture surfaces are not in contact and the tissue in the fracture gap cannot carry substantial loads. Under such circumstances the load at the fracture fixation device represents the load acting at the fracture when no fixation device would stabilize it. For the loading of the lower extremity, which is most often of interest for fracture healing studies, the ground reaction force (GRF) under the foot or hoof is measured by a force platform. The forces and moments measured at the ground can be compared to the loads measured at the fracture fixation device to find a correlation between the GRF and local load at the fracture level. This allows the estimation of the loads at the fracture for other experiments when the same fracture model and fixation device is used and the GRF is measured. Normally the loads at the ground and at the fracture level are different because the anatomical axis of the bone is not in the same direction as the load axis of the GRF and the internal muscle forces add loads to the external loads.

Another method is the calculation of the loads at the fracture level by using numerical musculo-skeletal models and the GRF. Such models are available for the human lower leg [4, 65, 66], and sheep [26] and rat hind limb [67]. Although the models are frequently validated by instrumented implants and GRF measurements, the numerical musculo-skeletal models can only calculate the loads at the fracture site with some inaccuracy (Figs. 3.6 and 3.7). They are unable to simulate the effect of, for example, pain or non-physiological loading after a fracture and operative stabilization.

The determination of loads at the fracture is in particular important for the analyses of mechano-biological rules in animal experiments when a direct measurement of the IFM is not possible. The load allows together with the stiffness of fracture fixation an estimated calculation of the IFM and IFS.

Normally, the mean vertical GRF during the stand-phase of the gait is used to characterize the loading of the fractured leg. However, there are other forces and moments acting at the contact between the leg and ground and there are, in addition to walking, numerous different activities with a significantly higher GRF possible during a fracture healing study. It remains unknown whether a large number of small load cycles or a small number of high load cycles are more influential on the bone healing process.

The GRF is most often standardized to the BW of the animal or patient to allow the comparison between various measurements. In humans, the GRF during gait is approximately 70–106% BW [68]. However it can be much higher when other activities like stair climbing, running, jumping or even unpredicted events like stumbling occur [69].

In quadruped animals, the GRF during gait is lower because the load is normally carried by one forelimb and one hindlimb at the same time. In mice, the mean peak GRF of the hindlimb is between 50% [70] and 63% [71] and for rats approximately 50% [71]. In sheep hindlimb loading of approximately 42% [72] was measured.

4.1.2.2   Forces and Moments at the Leg After Fracture Fixation

The GRF's of the uninjured legs increase after fracture or experimental fracture of the contralateral side when a fracture fixation is performed and the operated leg is only partially loaded. A significant unloading occurs in the injured leg after fracture treatment of patients because of the recommended partial unloading of the operated leg by the surgeon or because of postoperative pain and the application of the fracture fixation device [73]. These GRF's depend further on the form of activity, the individual injury and the behavior of the patients [73].

Likewise in experimental fracture healing studies (Fig. 4.8) [32, 51, 72] with animals, the postoperative loading is affected by the form of operation, type and

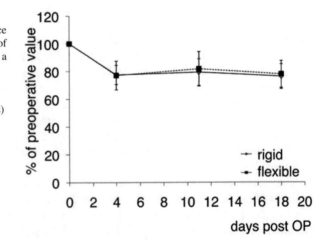

Fig. 4.8 Mean maximum GRF of the hind leg of mice treated with an osteotomy of the femur and stabilized by a rigid or flexible external fixator (see Fig. 4.6 left) (reprinted by permission from John Wiley and Sons) [51]

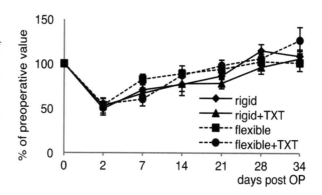

Fig. 4.9 Mean maximum GRF of the hind leg of rats treated with an osteotomy of the femur stabilized by two external fixators of different stiffness (rigid and flexible, see Fig. 4.6) (reprinted by permission from John Wiley and Sons) [44]. *TXT* rats suffered from additional thorax trauma

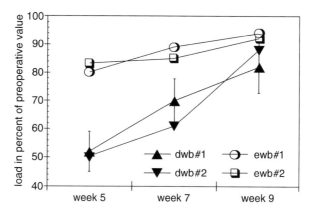

Fig. 4.10 Mean maximum GRF of the hind leg of sheep treated with an osteotomy of the tibia stabilized by an external fixator (reprinted by permission from Walters Kluwer Health) [74]. Two groups of sheep: dwb: delayed weight bearing, ewb: early weight bearing

stiffness of fracture stabilization, postoperative pain, activity of the animals and type of species. Normally, when the healing progresses, the immediate partial unloading postoperatively is followed by a steady recovery of the loading with time (Fig. 4.9). When this recovery does not occur, it is an indication of delayed healing (Fig. 4.10).

The postoperative GRF after operation and stable fixation of the fracture leg is reduced by 20–50% depending on the species and the possible restriction in loading. Typically, after the reduced loading of the limb, because of the operative intervention, the GRF increases with time and reaches preoperative values when the bone healing has bridged the fracture and allows full weight bearing. This is normally the time when clinically a bone healing is observed.

Whereas the difference in fixation stiffness in the shown mouse (Fig. 4.8) and rat experiments (Fig. 4.9) did not significantly affect the postoperative GRF, a very low fixation stiffness can cause a significant postoperative unloading of the limb [32] in sheep (Fig. 4.10).

4.1.3   Interfragmentary Movement

IFM can be measured in vivo for experimental animals [11, 29, 31, 32, 57] or for patients [2, 14, 75, 76]. However, this is technically demanding and was performed only in a few studies. Frequently, the IFM is approximately calculated from the stiffness of the fracture fixation and the estimated peak load during gait [26, 77, 78] at the beginning of the fracture healing process. This procedure to characterize the mechanical starting conditions has limitations because it assumes that the hematoma in the fracture gap and the soft tissue around it have no stiffness and the unknown individual loading can be realistically estimated.

The IFM changes with healing time (Fig. 4.11) because of alterations in loading of the injured extremity and the increasing stiffness of the fracture healing tissue (Callus formation) [15]. In undisturbed bone healing, the IFM decreases with time despite increasing loading of the fractured and stabilized bone [14], indicating that the stiffness of the healing bone increases more rapidly than the increase of the postoperative loading. These circumstances make it clear why it is extremely difficult to calculate the IFM at various healing time points because the bone loading and the stiffness of the healing fracture tissue are unknown.

The rate of decreasing IFM with healing time depends on several factors, including individual pain sensitivity and loading of the animal or patient [14], the individual biological healing capacity [57], the fracture gap size and the stability and characteristic of fixation (axial and shear stiffness of fixation) [29]. Figure 4.11 presents the course of osteotomy healing in sheep with different gap sizes and various IFM immediately postoperation [29]. All small gaps with small IFMs (groups A–C) displayed rapid healing with negligible IFM at the end of the study (nine weeks). Although the larger gaps in particular with a larger IFM showed a

Fig. 4.11 IFM as a function of healing time for six group of sheep (groups A–F) with an osteotomy of the tibia and stabilization with an external fixator that allowed adjustable osteotomy gap sizes and possible axial IFMs (reprinted by permission from John Wiley and Sons) [29]. A: gap 1 mm, IFM 0.1 mm, B: gap 1 mm, IFM 0.44 mm, C: gap 2 mm, IFM 0.2 mm, D: gap 2 mm, IFM 0.98 mm, E: gap 6 mm, IFM 0.38 mm, F: gap 6 mm, IFM 2.71 mm

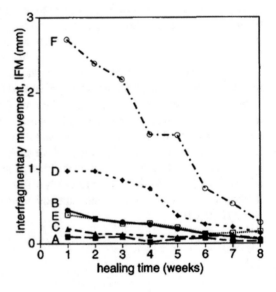

significant reduction in IFM because of larger callus formation, they exhibited delayed healing with significant IFMs at 9 weeks (groups D, F)

4.1.3.1 Monitoring Interfragmentary Movements

The number of studies that performed in vivo measurements of IFM is limited and those that monitored the IFM versus healing time are rare. Monitoring the IFM versus time enables the analysis of the progress during the various phases of bone healing. A change of the IFM indicates alterations in the stiffness of the healed fractures but does not necessarily correlate with the strength of the healing bone [79, 80].

Patient Monitoring

There are several methods to measure the 3D IFM applied in human tibia fractures stabilized by external fixators or braces [2, 15, 73, 81–84]. The displacement between bone screws implanted in both bone fragments under loading is measured by mechanical or optical systems and the displacement in the center of the fracture calculated from that data by a mathematical coordinate transformation. This allows the monitoring of the 3D IFM during the healing time (Fig. 4.12) A steady decrease of the IFM indicates a stiffening of the healing zone. This characteristic course of bone healing can however, be monitored without knowing the exact IFM in the center of the fracture, rather a value that correlates to it. Such a method is the measurement of the deformation of the external fixation system and is performed very early to monitor the fracture healing process [85]. This indirect method provides data that correlate well with the IFM and was successfully used in several studies (Fig. 4.13) [13, 86, 87].

When groups of patients are to be compared, for example, for clinical studies it is often necessary to normalize the data of the IFM because there is an extremely large variation of the IFM between individual patients even when the loading during the IFM measurement can be standardized very well. The main reason is that the fracture stabilization for each patient is individually very different depending on the type of injury and the strategy of the surgeon for the stabilization. This would lead to very large standard deviations of the mean value for each group and would make a statistical test unemployable. To overcome this problem a normalization of the IFM related to the first postoperative IFM representing 100% makes it possible to describe the course of the IFM versus the healing time for groups (Fig. 4.13). A steady decrease of the normalized IFM indicates the stiffening of the healing fracture. When a steady decrease of the IFM does not occur (Fig. 4.13) a delayed healing is indicated.

Fig. 4.12 Measurement of
interfragmentary movement
under loading of the lower
leg. A six-degree-of-
freedom goniometer system
is used to measure the
motion at the fracture gap
and a six-degree-of-freedom
load cell determines the load
applied to the foot (reprinted
by permission from
Elsevier) [2]. Both methods
together allow the
calculation of the stiffness of
the healing bone

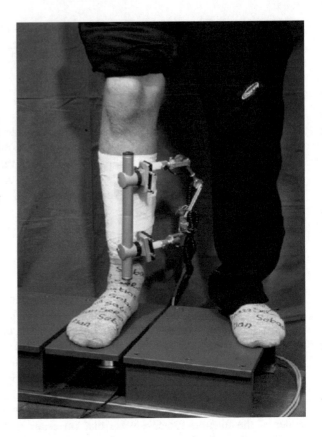

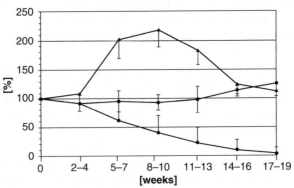

Fig. 4.13 Comparison of the normalized IFM's (IFM at the first postoperative measured time point
=100%) versus healing time for 100 patients with a tibia fracture stabilized by an external fixator
(reprinted by permission from Springer Nature) [86]. Normal healing characterized by (n =
92 patients, bottom curve) a steadily decreasing IFM, the "nonresponder group" (n = 5 patients,
middle curve) and patients with significant increase in IFM (delayed healing) (n = 3, top curve)
(mean values and standard deviations)

Monitoring Experimental Animals

Stürmer [88] studied the fracture healing in a transverse osteotomy model of the sheep tibia and monitored the axial and lateral IFM after external fixation. The course of the IFM versus time displayed a similar behavior to that of human patients but with a shorter time scale. During the first 2–3 weeks, there was a limited reduction in the IFM that was followed by a strong decrease in the IFM between 3 and 6 weeks. After 6 weeks the callus showed the beginning of bony bridging, which caused a reduction to very small IFMs. The experiment was performed with three groups of different initial postoperative IFMs between 0.5 and 2.5 mm resulting from different stiff external fixator arrangements. The larger IFMs displayed a more rapid reduction in value than the initially small IFMs, and led after 7 weeks to approximately the same IFMs in the late healing phase. These are very similar results to those found in the study of Claes et al. [29] (Fig. 4.11). In the latter study [29] different axial IFMs were applied for various fracture gap sizes. For small gap sizes of 1 and 2 mm, the initially larger IFMs in one group decreases more rapidly than the smaller IFMs in the other group with the same gap size. After approximately 7 weeks the IFM in both groups was similar (Fig. 4.11). Although the greater axial IFM stimulated a slightly larger callus formation, there was no statistical difference in postmortem bending stiffness between the groups. However, for the larger gap sizes, the reduction of the IFM after 7 weeks did not attain a value close to zero (see Fig. 4.11).

Klein et al. [79] monitored fracture healing in sheep tibia osteotomies stabilized in one group with an unreamed intramedullary nail and in the other group by an external fixator. During a period of eight weeks, the bending deflection and the torsional rotation were measured. Because of the lower torsional and bending stiffness of the intramedullary nail, the fracture stabilized with the nail displayed postoperatively significantly more bending deflection and axial rotation than the stabilization by an external fixator. Similar to the studies of Stürmer [88] and Claes et al. [29] the reduction in the IFM was more rapid for the fixation method with the higher initial IFM but after 5–7 weeks, the IFM was similar for both groups. However, this similarity in the IFMs in the late healing phase did not indicate a comparable quality of healing. The postmortem biomechanical testing of the healed bones revealed significantly greater torsional stiffness and strength for the more stable fixator group in comparison to the very flexible unreamed nail group. The histology showed a correspondingly greater callus width and a more mature callus formation in the external fixator group than in the nail group.

Park et al. [60] studied the fracture healing in a transverse osteotomy model of the rabbit tibia. The osteotomies were stabilized with a bilateral external fixator that was locket in one group and allowed an axial movement in the other group. The IFM was monitored during the healing time of 4 weeks. After approximately 2 weeks the IFM was reduced in the movement group to values comparable to the locked fixator group, which displayed only elastic deformation of the fixation system. While the healing was more rapid than for the sheep groups, it did display a similar characteristic in the decreasing IFM versus time. The movement group with IFM formed slightly more callus after 2 weeks than the locked fixator group, but this effect was absent after 4 weeks and there was no difference in the postmortem torsional stiffness or strength.

These results indicate that the monitoring of the IFM during the fracture healing process can provide information about the stiffening and progress of the callus formation in the early and middle phases (proliferation phase) of diaphyseal fracture healing, but does not allow conclusions to be drawn about the stiffness or strength in the late phase of healing when bony bridging of the fracture gap occurs. Whether the larger initial IFM stimulates a superior healing result is influenced more by the direction of the movement (axial versus shear) and the fracture gap size than by the capability of the callus formation to reduce the IFM in the proliferative phase of the healing process.

4.1.3.2 Interfragmentary Strain

The mechanically induced cellular reaction in the fracture healing zone, which drive tissue differentiation, is regulated by the tissue deformation (interfragmentary tissue strain, (IFS) resulting from IFM.

Perren and Cordey described one method to calculate the IFS by dividing the gap size by the amount of the IFM [89]. However, this calculation approximates the IFS only in the center of the cortical fracture zone under pure axial IFM. For other loading conditions, like bending, shear or torsion as well as for other fracture gap areas, the above-mentioned calculation of the IFS does not describe correctly the complex real IFS distribution in the fracture gap. A more realistic picture of the strain distribution in the fracture gap can be achieved when the anatomy and the loading of the fracture is simulated in a finite element model (FEM). Calculations from such a FEM can show the complex strain distribution in a fracture zone [90, 91] and allow a better correlation to cellular reactions and tissue differentiations. Such calculations allow discrimination between distortional and dilatational strain that are relevant for the cellular reaction [91, 92]. This is necessary when the mechanobiology of fracture healing is to be studied (see Sect. 6.1).

Because for most of the fracture healing studies no information about the strain distribution is known and only the IFM is described, merely an estimation of the IFS by dividing the gap size by the amount of IFM is possible.

4.1.3.3 Interfragmentary Movement, Tissue Strain and Tissue Differentiation

The effect of IFM on tissue differentiation depends mainly on the amount and direction of the IFM and the application of IFM in various healing phases. The knowledge about the effect of the IFM direction on bone healing is presented in Sect. 4.1.3.5 and about the effect of IFM during various healing phases in Sect. 5.2.

Most of the experimental studies are designed to compare different stable fracture fixations with the healing process from a clinical point of view. This chapter will focus on studies that investigated the tissue differentiation depending on the known initial axial postoperative IFM and the related tissue strain (IFS).

In the sheep diaphyseal osteotomy healing model initial IFM's of 0.2–1 mm (osteotomy gap size 1–3 mm) led to a rapid callus stimulation and bone healing by enchondral ossification [38, 93–96]. An IFM greater than 2 mm led to delayed healing with larger amounts of fibrous tissue and fibrocartilage [36, 95, 96]. Large IFM's in particular with large fracture gaps led to hypertrophic non-unions with predominantly fibrous tissue and cartilage [29, 36].

IFS calculated on the basis of the IFM and fracture gap size showed that a stimulation of callus formation and enchondral ossification occurs for strains of up to 40% [28, 31, 57, 96]. For IFS`s greater than 50%, no bone healing occurs and cartilage and fibrous tissue persist [33, 95, 97, 98]. In most of the experiments reported above, the initial IFM and IFS were reduced during the healing process because of the increased callus size and increasing stiffness and is very low when it comes to the callus bridging between the two fracture fragments. When the IFS is reduced below 6–7% a final bony bridging and healing can occur. In contrast when a larger IFS (greater than 10%) persists, healing cannot occur [99–101].

4.1.3.4 Interfragmentary Movement and Revascularization

Undisturbed bone healing requires sufficient vascularization of the fracture-healing zone. IFM (or the mathematically related interfragmentary tissue strain (IFS)) is an important factor affecting vascularization in the fracture-healing zone. Stable fixation with predominatly moderate interfragmentary compression movement has been demonstrated in numerous studies to stimulate vascularization and the bone healing process, whereas unstable fracture fixation delays these processes. Fracture fixation instability, particularly large shearing interfragmentary movement, can delay healing or even cause non-unions.

Good bone healing, as characterized by successful bone formation in the healing zone and the return of mechanical strength and stiffness, correlates with the degree of vascular perfusion in the fracture-healing zone [102] (Fig. 4.14).

IFM or IFS and the resulting vascularization are quantitatively described in only a limited number of studies [33, 103, 106–110].

Claes et al. [107] described the vascularization of the callus healing area in an ovine metatarsal osteotomy model with a 2 mm osteotomy gap after 9 weeks. A stiff ring-fixator that allowed an adjustable axial IFM of 0.2 mm or 1.0 mm under physiological loading of the extremity was applied to stabilize the osteotomy, resulting in an axial IFS of 9% or 32%, respectively. Significant difference was found between the distribution of small vessels (<20 µm diameter) and large vessels (>40 µm diameter). Large vessels were primarily found in the medullary cavity, whereas the greatest density of small vessels was present in the peripheral area of the periosteal callus. The lowest vessel density occurred in the cortical osteotomy gap

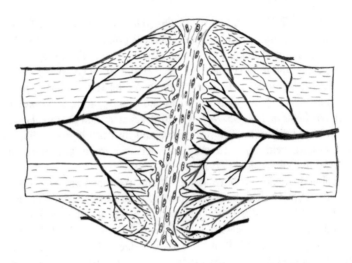

Fig. 4.14 Scheme of the vascularization in the callus healing zone (reprinted by permission from Elsevier) [103–105]. In particular large shear movements along the fracture line cause an avascular fibrous cartilage zone that cannot be bridged by vessels. Vessel formation stops at the border to the cartilage tissue with large tissue deformations

and in the medullary cavity. Generally, and particularly close to the periosteum, the number of vessels and quantity of new bone formation in the healing area was greater in the group with a smaller IFM and IFS. While a small amount of new bone formation was observed in the osteotomy gap of the cortex in the small IFS group, almost none was detectable in the large IFS group [103].

Similar results were reported in a study of Lienau et al. [106]. A 3 mm osteotomy gap of the sheep tibia was stabilized with either a rigid or a semi-rigid unilateral external fixator. The semi-rigid fixator group exhibited an axial IFM and a shearing IFM of 0.46 mm of 0.78 mm, respectively, in comparison to 0.32 mm and 0.52 mm, respectively, in the rigid group. The larger IFM found in the semi-rigid fixator group was associated with a reduced blood supply after both 2 and 9 weeks [106]. Similar to the study of Claes et al. [107], while large vessels were mainly found in the endosteal callus, small vessels were present most frequently in the periosteal area. Additionally as in Claes et al. [107], in both the early and late phases of the bone healing the small vessel density in the fibrous tissue area was greater for the rigid group than for the semi-rigid group. After 9 weeks, the callus formation in the rigid group displayed improved bone quality with greater stiffness in the healed bone relative to the semi-rigid group [103].

In a further study of Lienau et al. [109], various molecules related to blood vessel formation in the callus area were investigated under greatly different conditions of fracture stabilization than those described by Schell et al. [36]. Tibia osteotomies with a 3 mm gap in sheep where stabilized using a rigid or an extremely unstable fixator, resulting in a shearing IFM under loading of 0.7 and 9.5 mm in the rigid and unstable groups, respectively [36]. The axial compression 7 days postoperatively in the rigid and unstable groups was 0.4 mm (13% IFS) and 2.5 mm (83% IFS), respectively. The unstable fixation after 6 and 9 weeks caused significantly delayed bone healing or hypertrophic non-union, as demonstrated by a significantly lower torsional stiffness in comparison to the rigidly fixed group. The molecular biological analyses showed in the unstable group a delayed up-regulation or even a down-regulation of genes important for the vascularization in the healing area in comparison with the stable group. In some sheep with unstable fixation, hypertrophic non-union was found after 6 months [103, 109].

Callus formation and vascularization were investigated by Hente et al. [33] in a 2 mm osteotomy gap of the sheep tibia under controlled IFM. A triangular, wedge-shaped bone segment was cyclically tilted within a triangular bone defect with the center of rotation at the tip of the bone segment, thereby generating a gradient of compressive and tensile IFM and IFS ranging over 0–90% on opposing sides of the wedge. Different numbers of daily movement cycles were applied (0, 10, 1,000, or 10,000 cycles) and the bone formation in both the periphery and osteotomy gap together with the number and area of vessels in the osteotomy gap were determined. Cyclic compression stimulated significant bone formation, whereas this was only minimal on the tensile side. Increasing numbers of load cycles reduced the bone formation, with 10,000 load cycles per day preventing significant bone formation in the osteotomy gap. The optimal bone and vessel formation occurred with ten cycles per day on the compressive side. Peripheral callus formation increased with increasing IFS up to 55%. By contrast, bone formation in the osteotomy gap decreased with increasing IFS, becoming minimal for IFS >40%, with virtually no vessel formation in the gap under such conditions [103].

Wallace et al. [108] investigated the vascular response to fracture movement in a 2 mm transverse osteotomy of the ovine tibia stabilized using an instrumented external fixator. The sheep at 2 weeks post-surgery were assigned to two groups, in which, under physiological loading, the external fixator allowed an IFM of either 0.8 mm (40% IFS) or 1.3 mm IFM (65% IFS). Radioactive microspheres were used to determine the medullary and cortical blood flow. The region of interest was limited to 1 cm proximal and 1 cm distal to the osteotomy. At 6 weeks post-operation, greater callus formation, a greater failure torque and torsional stiffness, and a significantly higher blood flow were displayed by the larger IFM group in the cortex and medulla. These results contradict those of other studies that found better vascularization of the bone healing area for a smaller IFM and IFS. This may be explained by the fact that the blood flow in the periosteal callus was not determined and the method of quantifying the blood supply could not be directly compared to the other studies [103].

Despite being the dominant component (axial) in each of the previously discussed experiments, the IFM arising between the stabilized fracture fragments under loading was complex and three-dimensional. This limitation prevented the determination of the influence of the individual IFM modes (tension, compression, and shear) on vascularization and bone formation. Although a combination of tissue strain modes acts in the osteotomy gap of each of the above-mentioned studies, cyclic predominantly compressive strain appears to stimulate greater new vessel formation than cyclic tensile or shearing strain.

Recent studies reported the application of isolated IFM modes and investigated their effects on bone formation and vascularization [103]. These studies showed that new bone formation was stimulated at the surface of the sheep tibia by a lateral callus distraction process. The influence of subsequent cyclic loading in compression, tension, or shear on further bone formation was investigated [103]. It was shown that 120 loading cycles per day with a compressive amplitude of 0.6 mm (approximately 16% IFS at the beginning of stimulation) induces fourfold higher vascularization in the bone healing area compared to tensile or shearing stimulation of similar strain amplitudes. Consequently, the bone formation resulting under compression was twice that of tension or shear [103].

4.1.3.5 Directions of Interfragmentary Movement

As early as 1960, Pauwels [92] reported that a fracture display different tissue developments on the tension and compression side of the fracture healing zone under bending loads. He found greater callus formation with predominantly endochondral ossification on the compression side and smaller callus formation with mainly connective tissue on the tension side. These results are supported by an experimental study of Hente et al. [33]. A cyclic bending stimulation of an osteotomy of the sheep tibia induced up to 25 times greater periosteal callus formation on the compression compared with the distraction side [110].

Yamagishi and Yoshimura [59], demonstrated in a rabbit fracture model that compression movements led to better bone healing than shear movements that may result in delayed healing or non-union. Augat et al. [31] compared in a sheep tibia osteotomy experiment the bone healing under initially 1.5 mm axial or 1.5 mm shear movement. They found a significantly better bone healing under axial compression than under transversal shear movements. By contrast, Park et al. [60] found in a rabbit oblique fracture model better results for shear movements than for axial compression movements. The reason for this might be the different osteotomy orientation that does not lead to a pure shear movement.

Lienau et al. [106] investigated bone healing in a sheep osteotomy model stabilized with two different unilateral external fixators. Their study initially allowed larger translational shear IFM than axial compression under loading. They found that an increase in the initial translational shear movement from 0.52 mm in one experimental group to 0.78 mm in another group led to a decrease in mineralized bone and increased fibrous tissue [110].

Bishop et al. [34] applied cyclic compression and cyclic torsional shear movements to sheep translational osteotomies using an external fixation device and found that interfragmentary shear itself is not detrimental to bone healing.

A problem of all studies using elastic fixation systems is that the 3D stiffness of the fixation systems allows other uncontrolled deformations that are superimposed on the movements that intended be studied. The effect of such uncontrolled movements can distort the study results depending on the stiffness characteristics of the systems and might be responsible for differences between various studies.

To study the effect of isolated IFM directions on bone formation, novel devices were developed. These allows the isolated application of cyclic compressive, tensile or shearing movement [27, 110] to the bone-healing front. The study of Claes et al. [110] showed, that cyclic compression led to significantly more bone formation than cyclic tension or shear movement.

4.1.4   Fracture Gap Size and Geometry

Bhandari et al. [111] analyzed the healing of 200 human tibial fractures and indicated that one of the three main factors predicting complications in healing is the lack of cortical continuity (fracture gap) between the fracture ends. This effect of a fracture gap is supported by a number of experimental studies.

The quality of fracture healing outcome and the rate of healing decreases with increasing fracture gap size. This is shown for both, large animal models [29, 112, 113] and small animal models [40, 114, 115]. Large fracture gaps can create critical size defects that suppress healing and lead to non-unions [29, 113, 115, 116]. The biological capacity to grow bone on the existing bony fracture surface appears to be limited (Fig. 4.15) [113] independent of the biomechanical environment of the fracture. A critical size defect might be defined as exceeding 1.5–3 times the diameter of the bone [116].

Whereas the cortical fracture gap in a diaphyseal fracture most often remains unchanged until the peripheral callus bridging occurs (Fig. 4.16, FGc) the fracture

Fig. 4.15 Histological images of fracture healing in sheep tibia osteotomies (9 weeks p.o.). Longitudinal section, Trichrome-Goldner staining, magnification 5×. Left side 1 mm osteotomy (IFS: 31%) with bony bridging of the fracture and right side 6 mm osteotomy with remaining fracture gap (IFS: 31%). Calcified bone is shown in green and connective tissue in orange (reprinted by permission from Elsevier) [117]

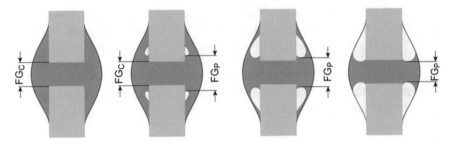

Fig. 4.16 Cortical fracture gap size of diaphyseal fractures (FGc) and changing fracture gap size at the peripheral callus wedges (FGp) versus time (from left to right)

gap at the level of the callus wedges is reduced with increasing callus formation (Fig. 4.16, FGp).

Besides the limited biological capacity for bone formation, the decreasing fracture gap size at the callus level however increases the IFS in the callus area when the IFM is not significantly reduced and can hinder bony bridging.

In the case of very stable fixation the mechanical stimulus for new bone formation is lacking and large gaps do not or very slowly close. The worst situation occurs for large fracture gaps with large IFM's. This combination is well known as an established non-union model [29, 99, 118].

In addition to the pap size, the geometry of the fracture or osteotomy can affect bone healing. A comparative study on oblique and transverse Osteotomies fixed with bony contact in canine tibiae [119] found better healing results for the transverse osteotomies even though the oblique osteotomy displayed the greater biologically active fracture surface. The possible reason could be the greater stability of the external fixation in transverse osteotomies with bony contact than for oblique osteotomies with bony contact. Another possible factor might be the tissue strain in the osteotomy, which is mainly compression strain for transverse osteotomies in contrast to additional shear strain in oblique osteotomies [120]. In contrast to this study a comparative study on rabbit tibia fractures that were stabilized by an axial sliding fixator showed greater callus formation and bone strength for the oblique fractures [60] than for the transverse fractures. IFM monitoring showed much smaller IFM's for the transverse fractures in comparison to the oblique fractures, which could explain the larger callus stimulation in the oblique fracture group and the corresponding greater strength of the healed bones.

More complex fractures are simulated by wedge shaped osteotomies [33] or comminuted fractures by triple wedge osteotomies creating several fragments [58].

The effect of the geometry of the fracture model on the bone healing process is very difficult to quantify because a number of other factors like fracture gap size, stiffness of fixation, surface characteristic and blood supply play a role. One important mechanobiological factor is the IFS. Shear strain is more pronounced in oblique fracture models than transverse fracture models and can lead to diminished bone healing [103, 120].

References

1. Duda, G.N., et al. 1998. A method to determine the 3-D stiffness of fracture fixation devices and its application to predict inter-fragmentary movement. *Journal of Biomechanics* 31 (3): 247–252.
2. Wehner, T., et al. 2010. Influence of the fixation stability on the healing time - A numerical study of a patient-specific fracture healing process. *Clinical Biomechanics* 25 (6): 606–612.
3. Grant, C.A., M. Schuetz, and D. Epari. 2015. Mechanical testing of internal fixation devices: A theoretical and practical examination of current methods. *Journal of Biomechanics* 48 (15): 3989–3994.
4. Heller, M.O., et al. 2001. Musculo-skeletal loading conditions at the hip during walking and stair climbing. *Journal of Biomechanics* 34 (7): 883–893.
5. Höntzsch, D., et al. 1993. Die begleitende Fibulaosteosynthese bei der kompletten Unterschenkelfraktur. *Trauma* 9: 110.
6. Claes, L. 2006. Biologie und biomechanik der osteosynthese und frakturheilung. *Orthopädie und Unfallchirurgie* 1 (4): 329–346.
7. Penzkofer, R., et al. 2009. Influence of intramedullary nail diameter and locking mode on the stability of tibial shaft fracture fixation. *Archives of Orthopaedic and Trauma Surgery* 129 (4): 525–531.
8. Schandelmaier, P., C. Krettek, and H. Tscherne. 1996. Biomechanical study of nine different tibia locking nails. *Journal of Orthopaedic Trauma* 10 (1): 37–44.
9. Horn, J., et al. 2009. Angle stable interlocking screws improve construct stability of intramedullary nailing of distal tibia fractures: a biomechanical study. *Injury* 40 (7): 767–771.
10. Claes, L. 2017. Mechanobiology of fracture healing. *Unfallchirurg* 120 (1): 13.
11. Kaspar, K., et al. 2005. Angle stable locking reduces interfragmentary movements and promotes healing after unreamed nailing. Study of a displaced osteotomy model in sheep tibiae. *The Journal of Bone and Joint Surgery. American Volume* 87 (9): 2028–2037.
12. Behrens, F., and K. Searls. 1986. External fixation of the Tibia. *Journal of Bone and Joint Surgery* 68: 246–254.
13. Claes, L. 1991. Measuring bone healing in osteosynthesis with external fixator using the Fraktometer FM 100. *Chirurg* 62 (4): 354–355.
14. Cunningham, J.L., M. Evans, and J. Kenwright. 1989. Measurement of fracture movement in patients treated with unilateral external skeletal fixation. *Journal of Biomedical Engineering* 11 (2): 118–122.
15. Gardner, T.N., et al. 1997. Dynamic interfragmentary motion in fractures during routine patient activity. *Clinical Orthopaedics* 336: 216–225.
16. Gasser, B., et al. 1990. Stiffness characteristics of the circular Ilizarov device as opposed to conventional external fixators. *Journal of Biomechanical Engineering* 112 (1): 15–21.
17. Gardner, T.N., and M. Evans. 1992. Relative stiffness, transverse displacement and dynamization in comparable external fixators. *Clinical biomechanics* 7: 231–239.
18. Schneider, E., et al. 1992. Zur biomechanik des ringfixateurs - beiträge einzelner strukturelemente. *Unfallchirurg* 95: 580–587.
19. Shefelbine, S.J., et al. 2005. Intact fibula improves fracture healing in a rat tibia osteotomy model. *Journal of Orthopaedic Research* 23 (2): 489–493.
20. Perren, S.M. 2002. Evolution of the internal fixation of long bone fractures. The scientific basis of biological internal fixation: choosing a new balance between stability and biology. *Journal of Bone and Joint Surgery. British Volume (London)* 84 (8): 1093–1110.
21. Stoffel, K., et al. 2003. Biomechanical testing of the LCP–how can stability in locked internal fixators be controlled? *Injury* 34 (2): 11–19.
22. Bottlang, M., et al. 2017. Dynamic stabilization of simple fractures with active plates delivers stronger healing than conventional compression plating. *Journal of Orthopaedic Trauma* 31 (2): 71–77.

23. Claes, L. 2011. Biomechanical principles and mechanobiologic aspects of flexible and locked plating. *Journal of Orthopaedic Trauma* 25 (1): 4–7.
24. Röderer, G., et al. 2014. Delayed bone healing following high tibial osteotomy related to increased implant stiffness in locked plating. *Injury* 45 (10): 1648–1652.
25. Ahmad, M., et al. 2007. Biomechanical testing of the locking compression plate: when does the distance between bone and implant significantly reduce construct stability? *Injury* 38 (3): 358–364.
26. Duda, G.N., et al. 1998. Analysis of inter-fragmentary movement as a function of musculo-skeletal loading conditions in sheep. *Journal of Biomechanics* 31 (3): 201–210.
27. Meyers, N., et al. 2017. Characterization of interfragmentary motion associated with common osteosynthesis devices for rat fracture healing studies. *PLoS One* 12 (4): e0176735.
28. Goodship, A.E., et al. 1993. The role of fixator frame stiffness in the control of fracture healing. An experimental study. *Journal of Biomechanics* 26 (9): 1027–1035.
29. Claes, L., et al. 1997. Influence of size and stability of the osteotomy gap on the success of fracture healing. *Journal of Orthopaedic Research* 15 (4): 577–584.
30. Wolf, S., et al. 1998. The effects of external mechanical stimulation on the healing of diaphyseal osteotomies fixed by flexible external fixation. *Clinical Biomechanics* 13 (5): 359–364.
31. Augat, P., et al. 2003. Shear movement at the fracture site delays healing in a diaphyseal fracture model. *Journal of Orthopaedic Research* 21 (6): 1011–1017.
32. Klein, P., et al. 2003. The initial phase of fracture healing is specifically sensitive to mechanical conditions. *Journal of Orthopaedic Research* 21 (4): 662–669.
33. Hente, R., et al. 2004. The influence of cyclic compression and distraction on the healing of experimental tibial fractures. *Journal of Orthopaedic Research* 22 (4): 709–715.
34. Bishop, N.E., et al. 2006. Shear does not necessarily inhibit bone healing. *Clinical Orthopaedics and Related Research* 443: 307–314.
35. Epari, D.R., et al. 2007. Timely fracture healing requires optimization of axial fixation stability. *Journal of Bone and Joint Surgery* 89 (7): 1575–1585.
36. Schell, H., et al. 2008. Mechanical induction of critically delayed bone healing in sheep: radiological and biomechanical results. *Journal of Biomechanics* 41 (14): 3066–3072.
37. Tufekci, P., et al. 2018. Early mechanical stimulation only permits timely bone healing in sheep. *Journal of Orthopaedic Research* 36 (6): 1790–1796.
38. Bottlang, M., et al. 2010. Far cortical locking can improve healing of fractures stabilized with locking plates. *The Journal of Bone and Joint Surgery. American Volume* 92 (7): 1652–1660.
39. Mark, H., et al. 2003. An external fixation method and device to study fracture healing in rats. *Acta Orthopaedica Scandinavica* 74 (4): 476–482.
40. ———. 2004. Effects of fracture fixation stability on ossification in healing fractures. *Clinical Orthopaedics and Related Research* 419: 245–250.
41. Mark, H., and B. Ryderik. 2005. Torsional stiffnessin healing fractures: Influence of ossification. *Acta Orthopaedica Scandinavica* 76 (3): 428–433.
42. Strube, P., et al. 2016. Influence of age and mechanical stability on the bone defect healing: Age reverses mechanical effects. *Bone* 42 (4): 758–764.
43. Claes, L., et al. 2009. Early dynamization by reduced fixation stiffness does not improve fracture healing in a rat femoral osteotomy model. *Journal of Orthopaedic Research* 27 (1): 22–27.
44. Recknagel, S., et al. 2011. Experimental blunt chest trauma impairs fracture healing in rats. *Journal of Orthopaedic Research* 29 (5): 734–739.
45. Glatt, V., et al. 2016. Reverse dynamization: A noval approach to bone healing. *Journal of the American Academy of Orthopaedic Surgeons* 24 (7): e60–e61.
46. Osagie-Clouard, L., et al. 2018. Biomechanics of two external fixator devices used in rat femoral fractures. *Journal of Orthopaedic Research* 37 (2): 283–298.
47. Harrison, L.J., et al. 2003. Controlled induction of a pseudarthrosis: A study using a rodent model. *Journal of Orthopaedic Trauma* 17 (1): 11–21.

48. Boerkel, J.D., et al. 2012. Effects of in vivo mechanical loading on large bone defect regeneration. *Journal of Orthopaedic Research* 30 (7): 1067–1075.
49. Nottebaert, M., et al. 2019. Omental angiogenic lipid fraction and bone repair. An experimental study in the rat. *Journal of Orthopaedic Research* 7 (2): 157–169.
50. Histing, T., et al. 2009. Ex vivo analysis of rotational stiffness of different osteosynthesis techniques in mouse femur fracture. *Journal of Orthopaedic Research* 27 (9): 1152–1156.
51. Rontgen, V., et al. 2010. Fracture healing in mice under controlled rigid and flexible conditions using an adjustable external fixator. *Journal of Orthopaedic Research* 28 (11): 1456–1462.
52. Epari, D.R., G.N. Duda, and M.S. Thompson. 2010. Mechanobiology of bone healing and regeneration: in vivo models. *Proceedings of the Institution of Mechanical Engineers* 224 (12): 1543–1553.
53. Chao, E.Y., et al. 1989. The effect of rigidity on fracture healing in external fixation. *Clinical Orthopaedics* 241: 24–35.
54. Wissing, H., and M. Stürmer. 1986. Untersuchungen zur Knochenregeneration nach Unterbrechung der medullären oder periostalen Strombahn bei verschiedenen Versuchstier-Species. *Hefte zur Unfallheilkunde* 181: 225–228.
55. Wissing, H., K.M. Stürmer, and G. Breidenstein. 1990. Die Wertigkeit verschiedener Versuchstierspezies für experimentelle Untersuchungen am Knochen. *Hefte zur Unfallheilkunde* 212: 479–488.
56. Nunamaker, D.M. 1998. Experimental models of fracture repair. *Clinical Orthopaedics and Related Research* 355: 56–65.
57. Claes, L.E., et al. 1995. Effect of dynamization on gap healing of diaphyseal fractures under external fixation. *Clinical biomechanics* 10 (5): 227–234.
58. Heitemeyer, U., et al. 1990. Significance of postoperative stability for bony reparation of comminuted fractures. An experimental study. *Archives of Orthopaedic and Trauma Surgery* 109 (3): 144–149.
59. Yamagishi, M., and Y. Yoshimura. 1955. The biomechanics of fracture healing. *Journal of Bone and Joint Surgery* 37: 1035–1068.
60. Park, S.H., et al. 1998. The influence of active shear or compressive motion on fracture-healing. *Journal of Bone and Joint Surgery* 80 (6): 868–878.
61. Histing, T., et al. 2011. Small animal bone healing models: standards, tips, and pitfalls results of a consensus meeting. *Bone* 49 (4): 591–599.
62. Willie, B., et al. 2009. Mechanical characterization of external fixator stiffness for a rat femoral fracture model. *Journal of Orthopaedic Research* 27 (5): 687–693.
63. Claes, L., et al. 2006. Moderate soft tissue trauma delays new bone formation only in the early phase of fracture healing. *Journal of Orthopaedic Research* 24 (6): 1178–1185.
64. Recknagel, S., et al. 2013. Conversion from external fixator to intramedullary nail causes a second hit and impairs fracture healing in a severe trauma model. *Journal of Orthopaedic Research* 31 (3): 465–471.
65. Duda, G., K. Eckert-Hübner, and L. Claes. 1997. Fracture gap movement as a function of musculo-skeletal loading conditions during gait. *Biomechanica* 97: 1.
66. Duda, G.N., E. Schneider, and E.Y. Chao. 1997. Internal forces and moments in the femur during walking. *Journal of Biomechanics* 30 (9): 933–941.
67. Wehner, T., et al. 2010. Internal forces and moments in the femur of the rat during gait. *Journal of Biomechanics* 43 (13): 2473–2479.
68. Bergmann, G., et al. 2001. Hip contact forces and gait patterns from routine activities. *Journal of Biomechanics* 34 (7): 859–871.
69. Bergmann, G., F. Graichen, and A. Rohlmann. 2004. Hip joint contact forces during stumbling. *Langenbeck's Archives of Surgery* 389 (1): 53–59.
70. Zumwalt, A.C., M. Hamrick, and D. Schmitt. 2006. Force plate for measuring the ground reaction forces in small animal locomotion. *Journal of Biomechanics* 39 (15): 2877–2881.
71. Clarke, K.A. 1995. Differential fore- and hindpaw force transmission in the walking rat. *Physiology & Behavior* 58 (3): 415–419.

72. Mora-Macias, J., et al. 2015. In vivo gait analysis during bone transport. *Annals of Biomedical Engineering* 43 (9): 2090–2100.
73. Duda, G.N.S., S. Sollmann, M. Hoffmann, J.E. Kassi, J.P. Khodadadyan, and C. Raschke. 2003. Interfragmentary movements in the early phase of healing in distraction and correction osteotomies stabilized with ring fixators. *Langenbeck's Archives of Surgery* 387 (11-12): 433–440.
74. Augat, P., et al. 1996. Early, full weightbearing with flexible fixation delays fracture healing. *Clinical Orthopaedics* 328: 194–202.
75. Gardner, T.N., et al. 1997. Temporal changes in dynamic inter fragmentary motion and callus formation in fractures. *Journal of Biomechanics* 30 (4): 315–321.
76. Duda, G.N., et al. 2002. Interfragmentary motion in tibial osteotomies stabilized with ring fixators. *Clinical Orthopaedics* 396: 163–172.
77. Bottlang, M., et al. 2009. Far cortical locking can reduce stiffness of locked plating constructs while retaining construct strength. *The Journal of Bone and Joint Surgery. American Volume* 91 (8): 1985–1994.
78. Gardner, T.N., M. Evans, and J. Kenwright. 1996. The influence of external fixators on fracture motion during simulated walking. *Medical Engineering & Physics* 18 (4): 305–313.
79. Klein, P., et al. 2004. Comparison of unreamed nailing and external fixation of tibial diastases–mechanical conditions during healing and biological outcome. *Journal of Orthopaedic Research* 22 (5): 1072–1078.
80. Chehade, M.J., et al. 1997. Clinical implications of stiffness and strength changes in fracture healing. *Journal of Bone and Joint Surgery. British Volume (London)* 79 (1): 9–12.
81. Claes, L.E., and J.L. Cunningham. 2009. Monitoring the mechanical properties of healing bone. *Clinical Orthopaedics and Related Research* 467 (8): 1964–1971.
82. Gardner, T.N., et al. 1994. Three-dimensional movement at externally fixated tibial fractures and osteotomies during normal patient function. *Clinical biomechanics* 9 (1): 51–59.
83. Sarmiento, A., et al. 1996. Effect of loading and fracture motions on diaphyseal tibial fractures. *Journal of Orthopaedic Research* 14 (1): 80–84.
84. Kempf, J., A. Grosse, and G. Beck. 1985. Closed locked intramedullary nailing. *Journal of Bone and Joint Surgery* 67: 709–720.
85. Burny, F.L. 1979. Strain gauge measurement of fracture healing. In *External fixation. The current state of art*, ed. A.F.J. Brooker and C.C. Edwards, 371–382. Baltimore: Williams and Wilkins.
86. Claes, L., et al. 2002. Monitoring and healing analysis of 100 tibial shaft fractures. *Langenbeck's Archives of Surgery* 387 (3-4): 146–152.
87. Richardson, J.B., J. Kenwright, and J.L. Cunningham. 1992. Fracture stiffness measurement in the assessment and management of tibial fractures. *Clinical biomechanics* 7 (2): 75–79.
88. Stürmer, K.M. 1988. Histologie und Biomechanik der Frakturheilung unter den Bedingungen des Fixateur externe. *Hefte zur Unfallheilkunde* 200: 233–243.
89. Perren, S.M., and J. Cordey. 1977. Tissue differences in fracture healing (author's transl). *Unfallheilkunde* 80 (5): 161–164.
90. DiGioia, A.M.I., E.J. Cheal, and W.C. Hayes. 1986. Three-dimensional strain fields in a uniform osteotomy gap. *Journal of Biomechanical Engineering* 108: 273–280.
91. Claes, L.E., and C.A. Heigele. 1999. Magnitudes of local stress and strain along bony surfaces predict the course and type of fracture healing. *Journal of Biomechanics* 32 (3): 255–266.
92. Pauwels, F. 1960. A new theory on the influence of mechanical stimuli on the differentiation of supporting tissue. The tenth contribution to the functional anatomy and causal morphology of the supporting structure. *Zeitschrift für Anatomie und Entwicklungsgeschichte* 121: 478–515.
93. Claes, L.E., et al. 1998. Effects of mechanical factors on the fracture healing process. *Clinical Orthopaedics and Related Research* 355: S132–S147.
94. Goodship, A.E., and J. Kenwright. 1985. The influence of induced micromovement upon the healing of experimental tibial fractures. *Journal of Bone and Joint Surgery* 67B (4): 650–655.

95. Kenwright, J., A. Goodship, and M. Evans. 1984. The influence of intermittent micromovement upon the healing of experimental fractures. *Orthopedics* 7 (3): 481–484.
96. Stürmer, K.M., T. Rack, and F. Kauer. 1990. Intravitale Bewegungsmessung bei der Frakturheilung. *Hefte zur Unfallheilkunde* 212: 489–498.
97. Hente, R.W., and S.M. Perren. 2021. Tissue deformation controlling fracture healing. *Journal of Biomechanics* 125: 110576.
98. Cheal, E.J., et al. 1991. Role of interfragmentary strain in fracture healing: ovine model of a healing osteotomy. *Journal of Orthopaedic Research* 9 (1): 131–142.
99. Cullinane, D.M., et al. 2002. Induction of a neoarthrosis by precisely controlled motion in an experimental mid-femoral defect. *Journal of Orthopaedic Research* 20 (3): 579–586.
100. Morgan, E.F., et al. 2010. Correlations between local strains and tissue phenotypes in an experimental model of skeletal healing. *Journal of Biomechanics* 43 (12): 2418–2424.
101. Palomares, K.T., et al. 2009. Mechanical stimulation alters tissue differentiation and molecular expression during bone healing. *Journal of Orthopaedic Research* 27 (9): 1123–1132.
102. Chidgey, L., et al. 1986. Vascular reorganization and return of rigidity in fracture healing. *Journal of Orthopaedic Research* 4 (2): 173–179.
103. Claes, L.E., and N. Meyers. 2019. The direction of tissue strain affects the neovascularization in the fracture-healing zone. *Medical Hypotheses* 137: 109537.
104. Rhinelander, F.W. 1974. Tibial blood supply in relation to fracture healing. *Clinical Orthopaedics* 105: 34–81.
105. Schweiberer, L., and R. Schenk. 1977. Histomorphologie und Vaskularisation der sekundären Knochenbruchheilung unter besonderer Berücksichtigung der Tibiaschaftfraktur. *Unfallheilkunde* 80: 275–286.
106. Lienau, J., et al. 2005. Initial vascularization and tissue differentiation are influenced by fixation stability. *Journal of Orthopaedic Research* 23 (3): 639–645.
107. Claes, L., K. Eckert-Hübner, and P. Augat. 2002. The effect of mechanical stability on local vascularization and tissue differentiation in callus healing. *Journal of Orthopaedic Research* 20 (5): 1099–1105.
108. Wallace, A.L., et al. 1994. The vascular response to fracture micromovement. *Clinical Orthopaedics* 301: 281–290.
109. Lienau, J., et al. 2009. Differential regulation of blood vessel formation between standard and delayed bone healing. *Journal of Orthopaedic Research* 27: 1133–1140.
110. Claes, L., et al. 2018. The mode of interfragmentary movement affects bone formation and revascularization after callus distraction. *PLoS One* 13 (8): e0202702.
111. Bhandari, M., et al. 2003. Predictors of reoperation following operative management of fractures of the tibial shaft. *Journal of Orthopaedic Trauma* 17 (5): 353–361.
112. Aro, H.T., and E.Y. Chao. 1993. Bone-healing patterns affected by loading, fracture fragment stability, fracture type, and fracture site compression. *Clinical Orthopaedics* 293: 8–17.
113. Markel, M.D., and J.J. Bogdanske. 1994. The effect of increasing gap width on localized densitometric changes within tibial ostectomies in a canine model. *Calcified Tissue International* 54 (2): 155–159.
114. Mehta, M., et al. 2012. In vivo tracking of segmental bone defect healing reveals that callus patterning is related to early mechanical stimuli. *European Cells & Materials* 24: 358–371.
115. Chaubey, A., et al. 2013. Structural and biomechanical responses of osseous healing: a novel murine nonunion model. *Journal of Orthopaedics and Traumatology* 14 (4): 247–257.
116. Garcia, P., et al. 2013. Rodent animal models of delayed bone healing and non-union formation: a comprehensive review. *European Cells & Materials* 26: 12.

117. Claes, L. 2021. Improvement of clinical fracture healing - What can be learned from mechano-biological research? *Journal of Biomechanics* 115: 110148.
118. Müller, J., R. Schenk, and H. Willenegger. 1968. Experimentelle untersuchungen über die entstehung reaktiver pseudarthrosen am hunderadius. *Helvetica Chirurgica Acta* 35: 301–308.
119. Aro, H.T., H.T. Wahner, and E.Y. Chao. 1991. Healing patterns of transverse and oblique osteotomies in the canine tibia under external fixation. *Journal of Orthopaedic Trauma* 5 (3): 351–364.
120. Steiner, M., et al. 2014. Numerical simulation of callus healing for optimization of fracture fixation stiffness. *PLoS One* 9 (7): e101370.

Chapter 5
Biomechanical Enhancement of Fracture Healing

Abstract There are several approaches for the biomechanical treatment of delayed unions or non-unions, including different methods of externally applied mechanical and physical signals. Five major methods are currently used: cyclic application of interfragmentary movements (IFM), low-magnitude high-frequency vibration (LMHFV), low-intensity pulsed ultrasound (LIPUS), low energy extracorporeal shockwave therapy (ESWT) and surgical dynamization of fractures. LIPUS and dynamization are clinically established methods, whereas ESWT is clinically seldom applied. LIPUS studies showed for non-operatively treated fractures significantly shorter healing times, but this effect could not be demonstrated for fractures treated by intramedullary nailing. Cyclically IFM at external fixators was previously and successfully applied and could significantly reduce the fracture healing time.

Some studies showed 50–70% better healing for non-unions treated by ESWT, but the quality of the studies was poor and this needs further investigation.

Dynamization can be performed by various techniques. Dynamization by telescoping can accelerate bone healing, while dynamization by destabilization of the fracture fixation in the late healing phase can reduce the time to full weight bearing.

Extremely rigid fixations can benefit from an early temporal, dynamization. When the fracture fixation was primarily very flexible, an alteration to a stable fixation should be performed as early as possible.

5.1   Biomechanical Enhancement of Fracture Healing

5.1.1   Externally Applied Methods

There are several approaches for the treatment of delayed unions or non-unions including different methods of externally applied mechanical or physical signals. Five major methods are currently used: cyclic application of interfragmentary movements (IFM), low-magnitude high-frequency vibration (LMHFV), low-intensity pulsed ultrasound (LIPUS), low energy extracorporeal shockwave therapy (ESWT) and surgical dynamization of fractures.

Cyclic application of IFM attempts to create an interfragmentary tissue strain (IFS) that is known to stimulate bone formation with a loading frequency that is similar to a physiological frequency during walking (approximately 1 Hz). LMHFV works with higher frequencies (20–90 Hz) but lower IFM amplitudes. In LIPUS, an acoustic signal with a much higher frequency (1 MHz) is applied while in ESWT, a sonic pulse broad frequency spectrum (16–20 MHz) is used [1]. The last two methods do not induce any IFM, rather aim to directly affect the cellular responds to improve the bone healing.

Studies that tested the effects of mechanical and physical signals on bone healing have a wide range of parameters of the applied stimulus. Frequency or strain rate of loading, number of load cycles, period of stimulation, energy applied, species, fracture model and fracture gap size are the most important parameters that vary between most of the published studies.

Therefore, a direct comparison of the study outcome is extremely difficult and conclusions have to been drawn carefully.

5.1.1.1 Cyclic Application of Interfragmentary Micromovements

Sometimes, the fracture fixation can lead to a strong immobilization of the fracture healing area that suppresses the bone formation necessary for successful healing. In such cases, an externally applied IFM can lead to a stimulus for new bone formation that enhances the bone healing process. The effect of such a treatment has been studied in a number of experimental and clinical studies.

Fracture healing in a sheep osteotomy model (fracture gap 3 mm) led to a stimulation of the bone formation and torsional stiffness of the healed bone when 500 cycles of axial IFM with an amplitude of 1 mm was applied daily [2]. The control group without external IFM application did not display any bony bridging of the fracture gap as a result of the very stiff stabilization of the osteotomy by a very rigid external fixator.

In another study of fracture healing stimulation by Goodship et al., the effect of the displacement rate of the IFM on bone healing was studied [3]. In an osteotomy model in sheep (3 mm fracture gap), stabilized by an external fixator, an IFM of 1 mm was cyclically applied with a displacement rate of 2, 40 or 400 mm/s (500 cycles/day, 0.5 Hz). The best healing results were observed after a stimulation of 40 mm/s.

In an osteotomy model of the sheep tibia (osteotomy gap 3 mm) and stabilization by an external fixator, the effect of a cyclic application of 0.2, 0.4, or 0.8 mm IFM (1 Hz, 1200 cycles/day) in the early repair phase (12 days p.o.) was studied [4]. After 6 weeks there was no significant difference between the stimulated groups and a non-stimulated control group. The control group already displayed good callus healing caused by the elastic deformation of the fixation due to weight bearing of the leg whereas the superimposed IFM cycles by external stimulation led only to a small amount of additional callus formation and bending stiffness of the healed bones [4].

A stimulation of fracture healing by cyclic bending was studied in an osteotomy (fracture gap 3 mm) model in sheep stabilized by an external fixator [5]. The bending moments applied to the external fixator led to an opening of the osteotomy of 0.2 or 0.8 mm. The cyclic IFM was applied with a frequency of 1, 5 and 10 Hz (500 load cycles/day). In comparison with the flexible fixed control group without external stimulation, the healing of the osteotomies in the stimulated groups was only slightly increased [5]. Flexible fixation in association with partial weight bearing of the leg already provides an adequate mechanical stimulus to induce good healing.

Another bending experiment showed that the bone formation by cyclic bending depends on the direction and number of load cycles of IFM [6]. An osteotomy model in sheep (2 mm fracture gap) was cyclically flexed with an opening of 1 mm on one side (tension) and a closing of the osteotomy of 1 mm on the other side of the tibia (compression). Ten or 1000 cycles per day were applied by a special external fixator and combined with a control group that was stabilized without externally applied IFM. There was more callus formation on the compression side than at the tensile side of the bone. Although 1000 cycles stimulated more callus formation than 10 cycles, the osteotomy could not heal under the higher number of enforced load cycles.

The better bone healing under compression could be confirmed in a recent study in sheep [7]. Cyclic compression, tension or shear movements were applied to the bone healing front on the sheep tibia. The bone formation and vascularization under compression was significantly better than under tension or shear movements.

In an experiment in sheep with a free (unloaded) bone segment the healing without IFM arising from functional loading could be studied [8]. Stimulation with 1 mm axial movement in an osteotomy gap of 3 mm (500 cycles/day at 1 Hz) led to significantly more bone formation and torsional strength of the stimulated osteotomy than the immobilized osteotomy.

The influence of the time of externally applied cyclic loading on fracture healing of bilateral rabbit tibia fractures stabilized by external fixation under a static 8 N fracture preload was studied by Wolf et al. [9]. After 21, 28, 35, 42 or 56 days a cyclic compression load of 40 N (55 Hz) was applied for 6 h/day in one leg till the end of the study after 8 weeks. The contralateral leg under constant static compression served as a control. The bones which were cyclically loaded after 6 weeks exhibited significantly greater torque to failure whereas early loading after 3 weeks appeared to diminish the torque to failure in comparison with bones under static compression [9].

The effect of cyclic compression and a distraction of 1 mm was tested in a rabbit osteotomy model with a fracture gap of 3 mm [10] and compared to a control group without external IFM. The cyclic deformation was applied with a frequency of 0.5 Hz and 50 cycles per day. There was more callus formation, a higher bone mineral content and greater stiffness of the healed bones in the cyclically stimulated groups in comparison to the control group regardless of the direction of the applied movement [10].

Gardner et al. investigated the effect of externally applied axial compression forces of different levels on the healing in mouse tibia fractures stabilized by intramedullary nailing under bony contact. One hundred loading cycles per day, 5 days per week for 2 weeks were applied after a delay of 0 or 4 days with 1 Hz frequency. An unloaded group served as control. Loading the fracture with a low load after 4 days improved fracture healing significantly as determined by increased callus strength, but this enhancement was lost as the load amplitude increased. Load initiation immediately following fracture inhibited healing, regardless of the load applied.

In the same animal model, a cyclic stimulation with a pause between each cycle did not improve the healing results compared with a control group without loading [11].

In a clinical study, the fracture healing in 80 patients with tibia fractures and stabilization by external fixators was investigated [12]. Two randomized groups were studied. One group received an externally applied IFM of 1 mm with a frequency of 0.5 Hz for 20 min each day and another group without stimulation served as a control. Every 2 weeks the bending stiffness of the healing bones was measured. When the bending stiffness attained 15 Nm per degree the bones were defined as clinically healed. The externally stimulated group achieved this healing state in average after 23 weeks, significantly earlier than the control group after a mean of 29 weeks [12].

From these studies it can be concluded that a significant mechanical stimulation and enhancement of fracture healing can be achieved by cyclically applied IFMs when the initial fracture fixation was very rigid and led to a suppression of the normal bone healing [2, 8, 9, 12]. IFM amplitudes that led initially to interfragmentary strains of up to 40% and stimulation frequencies of approximately 0.5–1 Hz appear to be sufficient to enhance the healing process. However, when the fracture fixation is flexible and provides in association with partial weight bearing an already adequate mechanical stimulus the effect of additional externally applied IFM is limited [4, 5].

An alternative to the very laborious application of cyclic IFMs to enhance the bone healing of very rigidly fixed fractures might be the dynamization of the fracture fixation to achieve more IFMs due to partial weight bearing and muscle activity of the broken bone.

5.1.1.2 Low-Magnitude High-Frequency Vibration (LMHFV)

LMHFV is a known successful anabolic strategy to increase bone mineral density [13] and was studied in regard to its possible application in fracture healing. Most of the in vitro cell culture studies showed that LMHFV is able to enhance mesenchymal stem cell and osteoblast proliferation, as well as osteogenic differentiation of mesenchymal stem cells whereas osteoclastogenic differentiation was inhibited [14]. The majority of studies both in small and large animal models demonstrated that LMHFV is able to accelerate bone formation in the fracture callus [14]. However, an increased callus volume does not necessarily lead to a biomechanically stronger bone healing [15] as shown in an LMHFV study in sheep.

Studies in mice showed that fracture healing was not significantly influenced by LMHFV with a frequency of 35 Hz but significantly reduced bone formation at 45 Hz stimulation [16]. In addition, the effect of LMHFV appears to be highly dependent on an animal estrogen status of the animals. In mouse models, LMHFV provoked negative effects on fracture healing in estrogen-competent animals, whereas it improved healing in estrogen -deficient animals [14]. No clinical studies with LMHFV are known to date.

5.1.1.3   Low Intensity Pulsed Ultrasound (LIPUS)

For more than 30 years, LIPUS has been used to stimulate fracture healing in particular in delayed unions and non-unions. The standard treatment, based on the most frequently used hardware, is a 20-min treatment per day of 1 MHz sine waves repeating at 1 KHz, mean intensity 30 mW/cm^2 and pulse width 200 µs [17]. There are a large number of experimental and clinical studies on the effect of LIPUS on the fracture healing outcome. The mode and mechanism of how LIPUS affects fracture healing is, meanwhile better understood. The LIPUS signal is transferred through the tissue to the fracture location and the cells in the healing region. Cells translate this physical signal to a biochemical signal via integrin mechano-receptors. This affects cyclo-oxygenese2 production, which in turn stimulates fracture healing. The results of numerous experimental and clinical studies on LIPUS effects in fracture healing are, however, unclear and appear to depend on various fracture conditions [17]. A meta-analysis of clinical studies [18] showed that from the large number of studies only six met the inclusion criteria. These studies showed a significantly shorter healing time for non-operatively treated fractures for the LIPUS group than for the control group. However, this effect could not be observed for fractures treated by intramedullary nailing.

5.1.1.4   Low Energy Extracorporeal Shockwave Therapy (ESWT)

ESWT is used i.e. to treat non-unions [19, 20] in patients as well as experimentally tested for the treatment of fractures of osteoporotic rats [21] and fresh fractures in sheep [22]. How ESWT influences the bone healing processes is not yet fully understood and remains the subject of discussion [21]. The first hypothesis was that ESWT causes microtraumata that stimulate a repair process [21]. More recent publications describe that neovascularization is induced through increased expression of angiogenic growth factors [21]. Furthermore, there is a direct activation of fibroblasts and subsequent transformation into osteoblasts [21]. There are a large number of studies regarding the effect of ESWT on fracture healing under very different stimulation conditions that appear to significantly affect the healing process. Different devices are used to create shock waves, including electro-hydraulic, piezo-electric and electro-magnetic [23]. Additionally, there is a wide range of energy applied [21] during the shock wave (0.0031–0.89 mj/mm^2) as well as different repetition rates and frequencies. The shock wave is characterized by a high peak pressure (500 bar), a short life cycle (10 ms), rapid pressure rise (10 ns), and a broad frequency spectrum (16–20 MHz) [1].

The wide range of the stimulation parameter may explain the very different results found in both, experimental and clinical studies. High energy flux intensities appear to have a detrimental effect on the bone healing [21, 22].

Reviews considering the effect of ESWT in clinical studies found that in approximately 50–70% of non-unions and delayed unions the bone healing could be stimulated [1, 24, 25]. However, the quality of most clinical studies was poor and further investigations are required.

5.1.2   Dynamization of Fractures

One widely used mechanical method to enhance fracture healing is the dynamization of fracture fixation. With this method the mechanical conditions of fracture fixation during the healing process are changed at a certain time point. However, there were conflicting results regarding the effect of the dynamization shown in clinical studies. Accelerated bone healing was found in some studies [26–31] of various types of tibial as well as femoral fractures. By contrast, other studies found no advantages of dynamization and did not recommend routine application of this method [32–35] while other authors even reported delay of bone healing [36–39]. One possible reason for the contradictory published results is the different conditions under which the dynamization was performed. The various types of fractures which were treated, the different kinds of fracture fixation, the different principles of dynamization applied, contact or gap healing, the different time points when the dynamization was initiated and the various outcome measures of fracture healing are important factors. This makes the comparison of these studies extremely difficult and a general conclusion as to whether or not a dynamization is a successful method for the enhancement of fracture healing cannot be concluded. To elucidate under which conditions a dynamization can be effective, the various methods of dynamization and the factors which influence the healing process should be analyzed, in particular, considering those experimental investigations which have more standardized fracture conditions and better control the effects of single factors on the healing outcome.

5.1.2.1   Methods of Dynamization

The term dynamization has been used in the literature for various types of mechanical changes during the fracture healing process. The most popular form of dynamization is axial telescoping, which allows one bone fragment to move in direction towards the other fragment. This is possible with intramedullary nails with locking screws that can be unlocked (Fig. 5.1) or with external fixators which, for example, have a system that allows an axial movement between the distal

Fig. 5.1 Dynamization by "telescoping" leads to a closure of the fracture gap, compression between the fragments and a reduction of interfragmentary movement (IFM) (adapted from [40]) (reprinted by permission from Springer Nature)

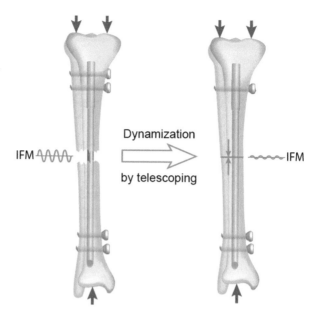

and proximal screw clamps. Under external loads and muscle forces the axial movement of one bone fragment relative to the other (telescoping) allows the closing of possible fracture gaps and contact and compression between the fragments. Frequently, the compression and force transmission between the fragments achieved by dynamization reduce the loading of the fixation and as a consequence the elastic deformation and cyclic IFM (Fig. 5.1) [41].

However, the telescoping system of external fixators does not always, guarantee the axial movement that is expected. Several studies have shown that under certain mechanical conditions the telescopic movement can be partly or totally suppress by "self-locking" [42, 43]. This can partly explain diverging clinical and experimental results with dynamization by external fixator telescoping systems.

Another form of dynamization is the increase of the IFM after a certain time of healing by increasing partial weight bearing of the injured leg. This leads for a given stiffness of a fracture fixation to a greater loading of the healing area and a higher IFM.

An additional dynamization (increase of IFM) can be achieved when at a certain time during the fracture treatment the stiffness of the fixation is reduced (by destabilization). This is possible, for example, when a fracture is stabilized by an external fixator with a double tube fixator body between the screw clamps or a larger number of screws. The removal of one tube (Fig. 5.2) or of screws reduces the stiffness of the fixation and consequently leads to a higher IFM under the same loading of the stabilized bone. The amount of IFM is dependent on the load applied to the bone and the stiffness of fixation and follows the cyclic behavior of the loading activity (Fig. 5.2).

Fig. 5.2 Dynamization by
destabilization of the
fracture fixation (reprinted
by permission from Springer
Nature) [40]

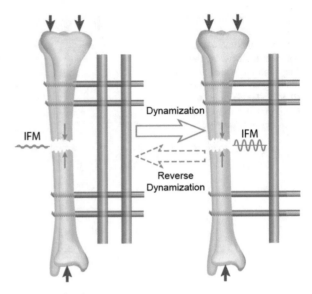

Some authors discussed whether reverse dynamization might help to enhance the healing process [44, 45]. In this procedure a flexible fixation of a fracture, with a relatively large IFM, is stiffened at a later phase of healing to improve the bone healing. Independent of the type of dynamization applied the question at which bone healing phase the dynamization should be initiated is one of the open key questions. Changes in the mechanical environment of the healing zone (dynamization) will have different effects depending at which bone healing phase they are applied. Therefore, knowledge about the healing phases of fractures as well as the sensibility of their tissue differentiation to mechanical signals (IFM) is important to understand the possible effects of dynamization.

5.1.2.2 Dynamization by Telescoping

Because it is well established that smaller fracture gaps heal more rapid than larger gaps [46–48], dynamization by telescoping (closing the gap) should, therefore, lead to an enhanced healing. In addition to the effect of closing the fracture gap, the compression of the fracture ends allows an increasing weight bearing [49], with a reduction of the loads acting at the fixation and consequently a smaller IFM at the fracture site [41].

The effect of telescoping was investigated in a bone healing study using a bilateral tibia osteotomy in dogs with an osteotomy gap of 2 mm [49] stabilized by an external fixator. In the dynamized group, the telescoping was allowed after 1 week whereas the control group remained locked with the 2 mm gap. After 6 weeks the torsional stiffness of the dynamized and healed bones were significantly higher than in the control group and there was significantly more bone formation in the osteotomy gaps of the dynamized bones in comparison with the control bones. This clearly demonstrated the enhancement of the healing by telescoping.

Another study of the same group [50] investigated the new bone formation after dynamization as a function of time. Bilateral tibia osteotomies in dogs with a 2 mm gap were stabilized by a rigid fixator. Dynamization was performed after 1 week in one extremity leading to a telescoping and contact of the fragments whereas the other leg remained locked. The dogs were sacrificed after 1 day, and 3, 5, 8 and 11 weeks after dynamization. Load bearing was greater on the dynamized limbs during standing for the first 5 weeks and during gait for the first 3 weeks after dynamization compared to the controls. The maximum periosteal callus size was attained more rapidly in the dynamized group and decreased after 12 weeks whereas there was no significant change on the control side. The dynamized side displayed significantly higher torsional stiffness at 6 weeks than did the control side. The authors concluded that early dynamization and telescoping accelerate callus formation and remodeling and provide greater mechanical stiffness during early stages of bone healing.

By contrast, a similar study, but with a smaller osteotomy gap size (0.8 mm) could not show an effect of dynamization [51]. The study stabilized bilateral tibia osteotomies in dogs by rigid external fixators that allowed dynamization by telescoping. The dynamization was performed in one leg after 15 days whereas the other leg was remained locked. After 90 days of healing there were no statistical differences in new bone formation or the maximum torque between the dynamized and control sides.

The main difference between the two mentioned studies was the osteotomy gap size (2 mm versus 0.8 mm). Whereas for the 2 mm gap size an improvement of the bone healing by dynamization could be shown, this advantage was absent for the small gap size. The reason for this might be that small fracture gaps under stable fixation heal very well anyway [48] and an additional effect of dynamization cannot make a significant contribution.

The effect of dynamization in a more complex fracture type was tested in a dog tibia osteotomy model creating a butterfly fragment that was stabilized by an intramedullary reamed interlocking nail [52]. Dynamization after 8 weeks by unlocking the nail led after 20 weeks of healing to a significantly greater torsional stiffness and strength of the healed bones than the locked control side. The authors concluded that although a positive effect of dynamization was demonstrated a dynamization was not recommended because of the risk when shortening unstable fractures.

Femoral fractures in rats stabilized by intramedullary locking nails after 6 weeks displayed greater callus formation and better mechanical properties and bone mineral content in bones that were dynamized after 20 days than in bones with static interlocking during the entire time of the investigation [53].

These studies show that a dynamization by telescoping accelerates the bone healing process in cases of simple transverse and short oblique fractures with remaining fracture gaps after primary stabilization and allow an earlier weight bearing. The positive effect of dynamization will be more pronounced in fractures with considerable gap sizes that can be closed by telescoping, than for small gaps and fractures with fragments which are already in contact [54]. This dynamization at very early bone healing phases appears not to be advantageous, whereas dynamization during the repair phase showed in all experimental studies an accelerating effect on bone healing.

Complex and comminuted fractures which do not allow the transfer of compression forces at the fracture zone should not be dynamized as long as no bony bridging is confirmed radiographically, because of the risk otherwise of shortening the bone.

5.1.2.3 Axial Cyclic Dynamization

The proposal for an early dynamization assume that in the inflammatory phase the cellular activity and differentiation is very sensitive to the mechanical environment and that this can trigger the entire healing process [44, 55–57].

In a rat fracture healing model fixed by an external fixator of adjustable stiffness (stiffness of the rigid fixator 74 N/mm, dynamized fixator 10 N/mm), an early dynamization 1 week after fracture did not improve the bone healing 5 weeks after the operation compared with rats which had a stable fixation during the entire time [58]. The early dynamization (larger IFM), in the inflammatory phase, led to a higher total callus volume but this callus was lower in mineral density, contained more cartilage [59] and displayed a lower flexural rigidity after 5 weeks compared to a stable fixation during the entire 5 weeks of the investigation [58].

The hypothesis of late dynamization is that the significant reduction of the IFM in the late phase of healing due to the large callus diameter and the stiffening of the callus tissue might not sufficiently stimulate callus maturation.

In particular, once the trabecular callus has formed, there is the risk that the maturation and remodeling process might stop at this stage, consequently inhibiting intracortical remodeling [60].

In the same rat model that was used for the early dynamization a late dynamization study was performed. Three or 4 weeks after fracture, in the late repair phase, dynamization was performed [58]. At the end of the study at 5 weeks, similar flexural rigidity was found of the bones that were dynamized after both, 3 and 4 weeks and the bones that had a stable fixation during the entire 5 weeks of the investigation. However, the dynamized bones displayed a better bone quality (modulus of elasticity) [61] and higher bone as well as lower cartilage content [59], suggesting an increased remodeling process and indicating a more advanced healing.

Similar results were found in a study with oblique osteotomies applied to canine tibial diaphysis. All osteotomies were stabilized by a very stable trilateral fixator (axial stiffness 661 N/mm) for the first 6 weeks and thereafter half of the osteotomies were dynamized by reducing the fixator, in the repair phase, to a six times more flexible unilateral configuration (114 N/mm) [62]. After 12 weeks the healed bones of the stably fixed and dynamized group displayed similar torsional stiffness, whereas the dynamized group showed significantly greater torsional strength and significantly more periosteal bone formation with a lower porosity, indicating a more advanced healing.

Utvag et al. [63] performed a study where rat femoral fractures were stabilized by a rigid reamed steel nail. In one group the rigid steel nail was replaced after 30 days by a flexible polyethylene nail (dynamization), whereas in a second group (control) the steel nail was replaced by the same type of steel nail than before. The dynamization, by higher flexibility of the polyethylene nail, increased the callus formation after 60- and 90-days but not the bending stiffness, whereas the maximal bending load was reduced in comparison to the control group. However, this operative procedure is not comparable with clinical dynamization techniques. The second operation interrupts the healing process and initiates a second inflammatory cascade which in particular under flexible conditions (polyethylene wire) impairs the healing process [64].

5.1.2.4 Reverse Dynamization

Reverse dynamization is an increase of the stiffness of fracture fixation at a certain time during the healing process. The hypothesis behind this procedure is that a low stiffness of the fracture fixation or no fixation (high IFM) in the early fracture healing phases causes higher IFM (Fig. 5.3), which stimulates the healing process [65] and larger callus formation leading to a greater stiffness and strength of the bone [44, 62, 66] during the repair phase. Greater stiffness of the fixation in the later phases should

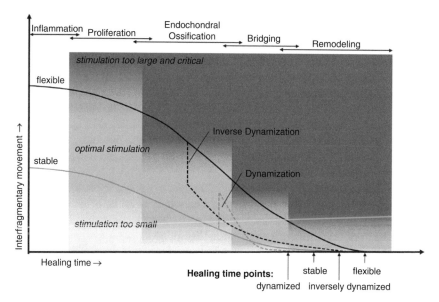

Fig. 5.3 Hypothetical effect of dynamization and reverse dynamization, by changing the fracture fixation stiffness, on the interfragmentary movement (IFM) and time of fracture healing. Green zone: postulated an optimal range of IFM for healing stimulation, yellow zone: small IFM with very low stimulation of bone healing, red zone: critical zone for IFM, risk for delayed healing and non-union

avoid critical IFM and the risk for a nonunion. In a recent study on rat femoral fractures the effect of reverse dynamization at different time points on bone healing was investigated [67]. The animal model was the same as the rat model for dynamization reported above with a rigid or flexible external fixator [61], with the only difference that the fixation was performed first with the flexible fixator which was changed to a rigid fixator after 3, 7 or 21 days. The animals were sacrificed after 5 weeks and compared to control groups with either rigid or flexible fixation for the entire 5 weeks. None of the healed bones with reverse dynamization displayed better mechanical or morphological results than the control group with a rigid fixation during the entire 5 weeks of treatment, whereas better results were observed than the group with flexible fixation for all 5 weeks.

Similar results were found in a study on tibia osteotomies in dogs [62] using an unilateral external fixator which was converted after 4 weeks to a rigid trilateral fixator in half of the bones. Increasing the stiffness of fixation after 4 weeks had no beneficial effect on osteotomy healing (torsional stiffness or maximum torque) and actually resulted in less bone in the periosteal region of the callus.

In a mouse fracture healing model on the tibia, it could be shown that a delayed fixation (1–4 days post op) produced a slightly greater callus volume after 10 days compared to an immediate fixation, this callus contained mainly more cartilage and not more bone. This led to similar failure loads under torsion independent of the time of reverse dynamization [57].

Tufekci et al. performed a study on a sheep model which isolates an experimental fracture from functional loading [8]. They compared a very rigidly fixed 3 mm tibia osteotomy with two groups that were externally stimulated by cyclical loading with a 1 mm amplitude. In a physiological-like group the stimulation was performed between days 5 and 21. Thereafter, the movements were decreased and stopped after 6 weeks. In the early stimulated group the cyclic loading was stopped after 3 weeks. Both stimulated groups displayed larger callus formation and torsional strength than the very rigidly fixed group, whereas there was no statistically significant difference between both stimulated groups. The authors concluded that it is sufficient to stimulate in the early phase of bone healing. However, their control group was extremely rigid and free of loads from load bearing and muscle forces and led to a suppression of bone formation.

These studies show that a flexible fixation in the early phase of bone healing and a later change to a more stable fixation might stimulate more cartilage in the callus formation but does not necessarily lead to a better bone healing than a sufficiently stable fixation of fractures during the entire time of healing.

Clinically a reverse dynamization should be performed in cases when obviously an unstable fixation was performed in the primary operation, that is, in an emergency situation of a polytraumatized patient. In such a case a correction of the unstable fixation to a stable fixation should be performed as early as possible.

The described beneficial effects of dynamization show a wide range of results. One of the reasons is that the results of these kinds of studies depend very much on the model conditions. When, for example, very rigid fixation is chosen for the control group, the stimulus for callus formation is small (Fig. 5.3) and a dynamization or externally applied cyclic IFM can improve the bone healing by stimulating larger callus formation [2, 8].

When the control group is stabilized by a suitable flexible fixation, it already develops a sufficient callus, and a superimposed IFM by dynamization or externally applied cyclic loads cannot significantly improve the bone healing [4].

Furthermore, the time point and period of dynamization is important. When, for example, the final end point of the study is very late the accelerating effect of dynamization, which could be detectable in an earlier healing phase, might be absent in the later phase.

Besides all the limitations due to various model conditions the described studies have shown, however, that when the chosen postoperative fracture fixation stability is unsuitable, a dynamization can be beneficial for the bone healing process.

Clinically, dynamization can be performed by various techniques. There is evidence that dynamization by telescoping in the repair phase of healing can accelerate bone healing and allow earlier weight bearing, in particular for simple transverse and short oblique fractures with large remaining fracture gaps [54]. Comminuted fractures that do not allow load transfer between bone fragments should not be dynamized until there is bony bridging as confirmed radiographically. Dynamization by destabilization of the fracture fixation at the transition from the repair phase to the remodeling phase or just after bony bridging of the peripheral callus appear to enhance the remodeling process and can reduce the time to full weight bearing [54]. For extremely rigid fixations a dynamization in the early phase of bone healing can be beneficial because of the stimulation of bone formation. When the fracture fixation is primarily very flexible a transfer to a stable fixation should be performed as early as possible.

References

1. Zelle, B.A., et al. 2010. Extracorporeal shock wave therapy: current evidence. *Journal of Orthopaedic Trauma* 24 (1): 66–70.
2. Goodship, A.E., and J. Kenwright. 1985. The influence of induced micromovement upon the healing of experimental tibial fractures. *Journal of Bone and Joint Surgery* 67B (4): 650–655.
3. Goodship, A.E., J.L. Cunningham, and J. Kenwright. 1998. Strain rate and timing of stimulation in mechanical modulation of fracture healing. *Clinical Orthopaedics* 355: 105–115.
4. Wolf, S., et al. 1998. The effects of external mechanical stimulation on the healing of diaphyseal osteotomies fixed by flexible external fixation. *Clinical biomechanics* 13 (4-5): 359–364.
5. Augat, P., et al. 2001. Mechanical stimulation by external application of cyclic tensile strains does not effectively enhance bone healing. *Journal of Orthopaedic Trauma* 15 (1): 54–60.
6. Hente, R., et al. 2004. The influence of cyclic compression and distraction on the healing of experimental tibial fractures. *Journal of Orthopaedic Research* 22 (4): 709–715.
7. Claes, L., et al. 2018. The mode of interfragmentary movement affects bone formation and revascularization after callus distraction. *PLoS One* 13 (8): e0202702.
8. Tufekci, P., et al. 2018. Early mechanical stimulation only permits timely bone healing in sheep. *Journal of Orthopaedic Research* 36 (6): 1790–1796.
9. Wolf, J.W., Jr., et al. 1981. Comparison of cyclic loading versus constant compression in the treatment of long-bone fractures in rabbits. *Journal of Bone and Joint Surgery* 63 (5): 805–810.

10. Matsushita, T., and T. Kurokawa. 1998. Comparison of cyclic compression, cyclic distraction and rigid fixation. Bone healing in rabbits. *Acta Orthopaedica Scandinavica* 69 (1): 95–98.
11. Gardner, M.J., et al. 2008. Pause insertions during cyclic in vivo loading affect bone healing. *Clinical Orthopaedics and Related Research* 466 (5): 1232–1238.
12. Kenwright, J., et al. 1991. Axial movement and tibial fractures. A controlled randomised trial of treatment. *Journal of Bone and Joint Surgery* 73 (4): 654–659.
13. Rubin, C., et al. 2004. Prevention of postmenopausal bone loss by a low-magnitude, high-frequency mechanical stimuli: a clinical trial assessing compliance, efficacy, and safety. *Journal of Bone and Mineral Research* 19 (3): 343–351.
14. Steppe, L., et al. 2020. Influence of low-magnitude high-frequency vibration on bone cells and bone regeneration. *Frontiers in Bioengineering and Biotechnology* 8: 595139.
15. Wolf, S., et al. 2001. Effects of high-frequency, low-magnitude mechanical stimulus on bone healing. *Clinical Orthopaedics* 385: 192–198.
16. Wehrle, E., et al. 2014. Distinct frequency dependent effects of whole-body vibration on non-fractured bone and fracture healing in mice. *Journal of Orthopaedic Research* 32 (8): 1006–1013.
17. Claes, L., and B. Willie. 2007. The enhancement of bone regeneration by ultrasound. *Progress in Biophysics and Molecular Biology* 93 (1-3): 384–398.
18. Busse, J.W., et al. 2002. The effect of low-intensity pulsed ultrasound therapy on time to fracture healing: a meta-analysis. *CMAJ* 166 (4): 437–441.
19. Schleberger, R., and T. Senge. 1992. Non-invasive treatment of long-bone pseudarthrosis by shock waves (ESWL). *Archives of Orthopaedic and Trauma Surgery* 111 (4): 224–227.
20. Valchanou, V.D., and P. Michailov. 1991. High energy shock waves in the treatment of delayed and nonunion of fractures. *International Orthopaedics* 15 (3): 181–184.
21. Mackert, G.A., et al. 2017. Low-energy extracorporeal shockwave therapy (ESWT) improves metaphyseal fracture healing in an osteoporotic rat model. *PLoS One* 12 (12): e0189356.
22. Augat, P., L. Claes, and G. Suger. 1995. In vivo effect of shock-waves on the healing of fractured bone. *Clinical Biomechanics* 10 (7): 374–378.
23. Auersperg, V., and K. Trieb. 2020. Extracorporeal shock wave therapy: an update. *EFORT Open Reviews* 5 (10): 584–592.
24. Rompe, J.D., et al. 1997. Extracorporeal shockwave treatment of delayed bone healing. A critical assessment. *Unfallchirurg* 100 (10): 845–849.
25. Willems, A., O.P. van der Jagt, and D.E. Meuffels. 2019. Extracorporeal shock wave treatment for delayed union and nonunion fractures: a systematic review. *Journal of Orthopaedic Trauma* 33 (2): 97–103.
26. Acker, J.H., C. Murphy, and R. D'Ambrosia. 1985. Treatment of fractures of the femur with the Grosse-Kempf rod. *Orthopedics* 8 (11): 1393–1401.
27. Foxworthy, M., and R.M. Pringle. 1995. Dynamization timing and its effect on bone healing when using the orthofix dynamic axial fixator. *Injury* 26 (2): 117–119.
28. Basumallick, M.N., and A. Bandopadhyay. 2002. Effect of dynamization in open interlocking nailing of femoral fractures. A prospective randomized comparative study of 50 cases with a 2-year follow-up. *Acta Orthopaedica Belgica* 68 (1): 42–48.
29. Johnson, K.D. 1985. Indications, instrumentation, and experience with locked tibial nails. *Orthopedics* 8 (11): 1377–1383.
30. Kempf, J., A. Grosse, and G. Beck. 1985. Closed locked intramedullary nailing. *Journal of Bone and Joint Surgery* 67: 709–720.
31. Klemm, K.W. 1986. Treatment of infected pseudarthrosis of the femur and tibia with an interlocking nail. *Clinical Orthopaedics and Related Research* 212: 174–181.
32. Brumback, R.J., et al. 1988. Intramedullary nailing of femoral shaft fractures. Part II: Fracture-healing with static interlocking fixation. *The Journal of Bone and Joint Surgery. American Volume* 70 (10): 1453–1462.
33. Vecsei, V., and J. Häupl. 1989. The value of dynamic adjustment in locking intramedullary nailing. *Aktuelle Traumatologie* 19 (4): 162–168.

34. Melendez, E.M., and C. Colon. 1989. Treatment of open tibial fractures with the Orthofix fixator. *Clinical Orthopaedics and Related Research* 241: 224–230.
35. Wiss, D.A. 1986. Flexible medullary nailing of acute tibial shaft fractures. *Clinical Orthopaedics* 212: 122–132.
36. Noordeen, M.H., et al. 1995. Cyclical micromovement and fracture healing. *Journal of Bone and Joint Surgery. British Volume (London)* 77 (4): 645–648.
37. Tigani, D., et al. 2005. Interlocking nail for femoral shaft fractures: is dynamization always necessary? *International Orthopaedics* 29 (2): 101–104.
38. Wu, C.C. 1997. The effect of dynamization on slowing the healing of femur shaft fractures after interlocking nailing. *The Journal of Trauma* 43 (2): 263–267.
39. Wu, C.C., and W.J. Chen. 1997. Healing of 56 segmental femoral shaft fractures after locked nailing. Poor results of dynamization. *Acta Orthopaedica Scandinavica* 68 (6): 537–540.
40. Claes, L. 2018. Dynamization of fracture fixation : timing and methods. *Unfallchirurg* 121 (1): 3–9.
41. Richardson, J.B., et al. 1995. Dynamisation of tibial fractures. *Journal of Bone and Joint Surgery* 77 (3): 412–416.
42. Ralston, J.L., et al. 1990. Mechanical analysis of the factors affecting dynamization of the orthofix dynamic axial fixator. *Journal of Orthopaedic Trauma* 4 (4): 449–457.
43. Howard, C.B., et al. 1999. Do axial dynamic fixators really produce axial dynamization? *Injury* 30 (1): 25–30.
44. Epari, D.R., et al. 2013. A case for optimising fracture healing through inverse dynamization. *Medical Hypotheses* 81 (2): 225–227.
45. Kenwright, J., and T. Gardner. 1998. Mechanical influences on tibial fracture healing. *Clinical Orthopaedics and Related Research* 355: S179–S190.
46. Aro, H.T., and E.Y. Chao. 1993. Bone-healing patterns affected by loading, fracture fragment stability, fracture type, and fracture site compression. *Clinical Orthopaedics* 293: 8–17.
47. Bhandari, M., et al. 2003. Predictors of reoperation following operative management of fractures of the tibial shaft. *Journal of Orthopaedic Trauma* 17 (5): 353–361.
48. Claes, L., et al. 1997. Influence of size and stability of the osteotomy gap on the success of fracture healing. *Journal of Orthopaedic Research* 15 (4): 577–584.
49. Egger, E.L., et al. 1993. Effects of axial dynamization on bone healing. *The Journal of Trauma* 34 (2): 185–192.
50. Larsson, S., et al. 2001. Effect of early axial dynamization on tibial bone healing: a study in dogs. *Clinical Orthopaedics and Related Research* 388: 240–251.
51. Aro, H.T., et al. 1990. The effects of physiologic dynamic compression on bone healing under external fixation. *Clinical Orthopaedics* 256: 260–273.
52. Georgiadis, G.M., G.J. Minster, and B.R. Moed. 1990. Effects of dynamization after interlocking tibial nailing: an experimental study in dogs. *Journal of Orthopaedic Trauma* 4 (3): 323–330.
53. Utvag, S.E., D.B. Rindal, and O. Reikeras. 1999. Effects of torsional rigidity on fracture healing: strength and mineralization in rat femora. *Journal of Orthopaedic Trauma* 13 (3): 212–219.
54. Claes, L. 2021. Improvement of clinical fracture healing - What can be learned from mechano-biological research? *Journal of Biomechanics* 115: 110148.
55. Chao, E.Y., N. Inoue, and J. Elias. 1998. Enhancement of fracture healing by mechanical and surgical intervention. *Clinical Orthopaedics* 355: 163–178.
56. Goodship, A.E., T.J. Lawes, and C.T. Rubin. 2009. Low-magnitude high-frequency mechanical signals accelerate and augment endochondral bone repair: preliminary evidence of efficacy. *Journal of Orthopaedic Research* 27 (7): 922–930.
57. Miclau, T., et al. 2007. Effects of delayed stabilization on fracture healing. *Journal of Orthopaedic Research* 25 (12): 1552–1558.

58. Claes, L., et al. 2009. Early dynamization by reduced fixation stiffness does not improve fracture healing in a rat femoral osteotomy model. *Journal of Orthopaedic Research* 27 (1): 22–27.
59. Willie, B.M., et al. 2011. Temporal variation in fixation stiffness affects healing by differential cartilage formation in a rat osteotomy model. *Clinical Orthopaedics and Related Research* 469: 3094–3101.
60. Hente, R., et al. 1999. Fracture healing of the sheep tibia treated using a unilateral external fixator. Comparison of static and dynamic fixation. *Injury* 30 (1): A44–A51.
61. Claes, L., et al. 2011. Late dynamization by reduced fixation stiffness enhances fracture healing in a rat femoral osteotomy model. *Journal of Orthopaedic Trauma* 25 (3): 169–174.
62. Egger, E.L., et al. 1988. Effects of destabilizing rigid external fixation on healing of unstable canine osteotomies. *Orthopaedic Research Society* 1988: 302.
63. Utvag, S.E., et al. 2001. Influence of flexible nailing in the later phase of fracture healing: strength and mineralization in rat femora. *Journal of Orthopaedic Science* 6 (6): 576–584.
64. Recknagel, S., et al. 2013. Conversion from external fixator to intramedullary nail causes a second hit and impairs fracture healing in a severe trauma model. *Journal of Orthopaedic Research* 31 (3): 465–471.
65. Coutts, R.D., et al. 1982. The effect of delayed internal fixation on healing of the osteotomized dog radius. *Clinical Orthopaedics and Related Research* 163: 254–260.
66. van Niekerk, J.L., et al. 1987. Duration of fracture healing after early versus delayed internal fixation of fractures of the femoral shaft. *Injury* 18 (2): 120–122.
67. Bartnikowski, N., et al. 2017. Modulation of fixation stiffness from flexible to stiff in a rat model of bone healing. *Acta Orthopaedica* 88 (2): 217–222.

Chapter 6
Mechanobiological Hypotheses, Numerical Models and Their Application to the Improvement of Clinical Fracture Treatment

Abstract The first tissue differentiation hypotheses were based on the observation that mesenchymal tissue that was stretched could develop into intramembranous bone, whereas mesenchymal tissue under hydrostatic pressure led to endochondral ossification.

Later, this observation was developed to a quantitative tissue differentiation hypothesis. Whereas these hypotheses are based on elastic, solid and isotropic material properties, others are based on biphasic material consisting of a fluid and a solid phase.

Depending on the tissue differentiation hypothesis used, the mechanical stimulus can be strain energy density, distortional strain, dilatational strain, fluid flow or combinations of these. The chapter focuses on a method using the fuzzy logic method. Using this model, it was finally possible to define optimal axial and shear stiffness values for fracture fixations to obtain a rapid bone healing.

The optimal stability was found to be an axial stiffness of between 1000 and 2500 N/mm and a shear stiffness larger than 300 N/mm. Larger diameter intramedullary nails fulfilled that criteria better than unreamed titanium nails with a small diameter and flexible external fixators. Conventional compression plates and interlocking plates often have an extremely high axial stiffness that suppresses the bone healing. To improve the fixation stiffness, several technical and surgical changes are proposed.

6.1 Mechanobiological Hypotheses of Tissue Differentiation in Fracture Healing

The first well accepted tissue-differentiation hypothesis was published by Pauwels [1]. He hypothesized that the cells embedded in the extracellular matrix of the fracture healing tissue can only sense two forms of deformation, pure shape distortion and pure volumetric changes (Fig. 6.1). This can lead to two kinds of cellular reaction. When the cells sense distortion under compression or tension (distortional strain), they will produce fibrous tissue, whereas a volumetric change under a

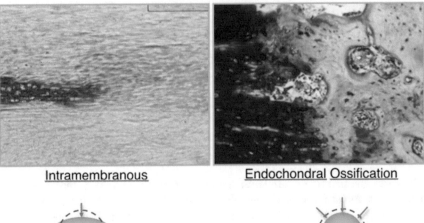

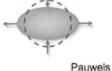

Fig. 6.1 Based on Pauwels' hypothesis, small distortional strain leads to intramembranous bone formation (left) and dilatational strain (hydrostatic pressure) to endochondral ossification (right)

hydrostatic pressure (dilatational strain) will lead to the production of cartilaginous tissue. Bone can develop from fibrous tissue or cartilaginous tissue when the local biomechanical conditions are favorable. For small deformations, bone can differentiate from fibrous tissue via intramembranous bone formation or from cartilaginous tissue by endochondral ossification (Fig. 6.2a)

Unfortunately, neither the distortional strain nor the hydrostatic pressure can be measured in bone healing tissue in vivo and, therefore, the important stimuli which drive the tissue differentiation are not known and tissue differentiation hypotheses cannot be proved in comparison with in vivo bone healing studies.

To overcome this problem, numerical models have been developed which calculate the biomechanical conditions in the bone healing area (tissue strain and/or hydrostatic pressure).

Perren [2] defined the simplest numerical model of tissue differentiation. In his interfragmentary strain (IFS) [2] theory, he proposed that the type of tissue which can differentiate in a fracture gap depends on the resilience against the acting axial IFS (IFS = axial IFM divided by the fracture gap size) without rupture. If it can resist the strain, it reduces the strain and stabilizes the fracture zone by proliferation to a mechanically stronger tissue, which finally allows the formation of new bone and fracture healing. A large IFS impedes bone healing, because only granulation tissue and soft tissue can tolerate large IFS's [3].

Calculations from finite element models (FEM's) of the strain distribution in the fracture healing zone [4] however, showed that the strain field inside the fracture gap region is much too complex to be represented by pure axial IFS alone. This is one of

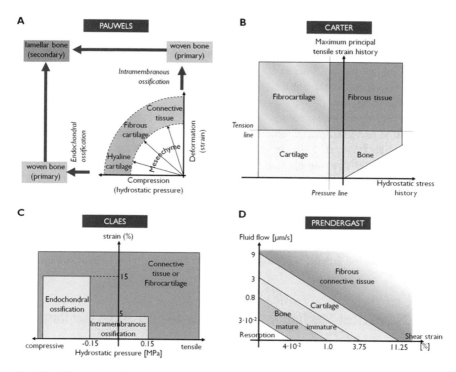

Fig. 6.2 Different hypotheses for tissue differentiation in bone healing: (**a**) Pauwels, (**b**) Carter, (**c**) Claes and Heigele, (**d**) Prendergast

the reasons why the IFS theory of Perren cannot explain various types of tissues, which typically proliferate in different areas of the fracture healing zone. In addition, this theory predicts a better healing when the fracture gap increases for a given IFM, because in consequence this leads to a decrease of the IFS. However, this is only valid for very small fracture gaps and not for larger gaps, where increasing gap sizes lead to an increasing delay of bone healing.

On the basis of Pauwels' theory, Carter [5] developed a semiquantitative tissue differentiation hypothesis, which proposed the hydrostatic pressure and the octahedral stress to be the determining stimuli for the tissue differentiation in a fracture healing zone [5–7]. High shear strain and/or positive hydrostatic pressure would lead to the proliferation of fibrous connective tissue, which can differentiate to bone depending on the vascularization and amount of shear stress and hydrostatic pressure (Fig. 6.2b). The prediction of which tissue will develop was described by an "osteogenic factor". Unfortunately, this index describes more or less a general behavior of the healing tissue and is not based on quantitative values for strain or hydrostatic pressure. Therefore, it cannot be used for numerical simulations of realistic fracture healing situations.

To overcome this problem Claes and Heigele [8] developed a tissue transformation hypothesis based on the hypothesis of Pauwels but with the aim to quantify the transition from one type of tissue to another. To achieve this aim, they calculated the strain and hydrostatic pressure distribution corresponding to histological images from three healing stages of a former ovine fracture healing experiment. The comparison of the strain and hydrostatic pressure distribution and the various tissue types in the histological images allowed the development of a quantitative description of the mechanical stimuli which caused a certain tissue transformation (Fig. 6.2c) [8, 9].

However, all of these static models simplify the situation in the callus by (linear) elastic, solid and isotropic material descriptions. For hard and more solid biological materials like bone, this simplification closely meets realistic material behavior. However, for softer materials like fibrous connective, cartilaginous or granulation tissue, which can frequently be divided into a solid and a fluid phase, this simplification is rather crude.

By contrast, Prendergast described the bone healing tissue as a biphasic (poroelastic) material consisting of a fluid and a solid phase. He then hypothesized that the tissue differentiation (Fig. 6.2d) depends on the distortional strain of the solid phase and the flowing velocity of the fluid phase within the strained tissue [10].

6.2   Numerical Mechanobiological Models for Bone Healing

The above-described tissue differentiation rules can be used to numerically simulate the bone healing process by stepwise differentiation of tissues as a function of the mechanical stimuli. Normally, two dimensional (2D) or 3D FEM's of the fracture healing zone are developed, which under an assumed stiffness of fragment fixation and loading experience a deformation of the healing tissue (Fig. 6.3) and allow the calculation of the distribution of mechanical stimuli.

Fig. 6.3 Finite element model of a diaphyseal bone with a fracture stabilized by an external fixator (represented by a spring system) (adapted and reprinted by permission from Springer) [11]

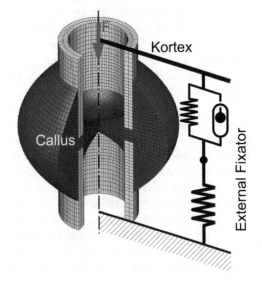

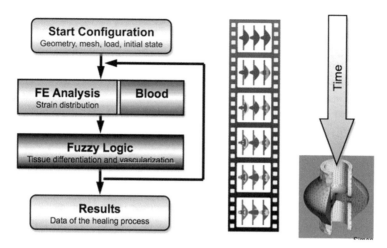

Fig. 6.4 Iterative procedure to simulate the stepwise changes in tissue differentiation during fracture healing (adapted and reprinted by permission from Springer) [11]

For each element in the fracture healing area, the initial distribution of tissue types and material properties and the mechanical stimuli are computed. On the basis of these stimuli, a tissue differentiation procedure modifies the tissue distribution and material properties depending on the differentiation rule chosen. In a next step, this change of material properties in turn changes the local mechanical behavior and, therefore, alters the mechanical stimuli, which again changes the tissue distribution. This iterative procedure is normally performed until the bony bridging of the fracture model has occurred (Fig. 6.4), indicating the final bone healing.

Depending on the tissue differentiation hypothesis, the mechanical stimulus can be strain energy density, distortional strain, dilatational strain, fluid flow or combinations of these. In addition to the mechanical stimuli, it is necessary to set some boundary conditions to realistically simulate the bone healing process. That is, following the rules, small mechanical stimuli allow a tissue differentiation from hematoma to bone at locations distant from the bone healing area, which is physiologically not possible. Therefore, for example boundary conditions that allow bone to form only in the neighborhood of an existing bone surface [8] are necessary, which is done by "if" (bone exists in the neighborhood) and "than" (bone can be formed) rules or by more complex rules. Ament and Hofer [12] introduced the fuzzy logic method to take additional and more complex rules for the tissue differentiation into consideration and allowed each finite element to also contain a mixture of the involved tissues (for example fibrous connective tissue, hyaline cartilage and bone).

The stimulatory strain energy combined with the fuzzi rules determine the change of the tissue composition after each iterative step.

In addition to the tissue differentiation rules described above, most of the numerical models include rules for revascularization of the healing area, because a blood supply is a prerequisite for all healing processes.

The influence of mechanical stimuli on tissue differentiation can be simulated on different levels. Whereas some authors modeled the healing process on the basis of the tissue types seen in the healing area, other authors attempted to model the behavior of cells that produce this tissue.

Depending on the goal that should be achieved (e.g., whether it should predict the phenomenological aspects on a cellular level or the macroscopic progress of fracture healing on an organ level), both approaches themselves and their combinations deliver important information. Cell-based models, however, have the great problem that there are numerous biological factors that are involved in cell activity and proliferation which are not well known and are very difficult to measure experimentally. Therefore, these models are limited by considerable uncertainties and need numerous decisive assumptions. Moreover, with an increasing number of unknown and assumed parameters, the validation becomes ever more complex.

Models that aim to allow predictions of the healing process in regard to the stability of the bone healing, like callus stiffness or reduction of the IFM, must allow a sufficient validation. The validation can be achieved by comparing the course of IFM during the healing progress with the course calculated by the simulation. This validation was performed for only a few models, with a notable difference found between the in vivo measurements and the calculated data [13, 14]. By contrast, the model developed in Ulm by Ament and Hofer [12], Simon et al. [15], Wehner et al. [16] and Steiner et al. [17] showed a good agreement between simulation and animal experimental results primarily for the IFM, but also secondarily for the progress of the tissue distribution over the healing process. To achieve this, Steiner et al. [17] improved the model by performing a model calibration based on four different load cases of sheep experimental data (i.e., small and large axial compression, translational, as well as torsional shear loading). The calibrated model was then able to realistically predict the healing process not only for the calibrated load cases,. but also for additional axial loading scenarios as well as bending, which showed the improvement of validity of the model. Therefore, the Ulm healing algorithm for the first time showed the applicability of healing simulation for various loading regimes of sheep experiments.

With the calibrated model, it was then possible to identify adverse mechanical conditions for fracture healing, and it was found that axial compression has more beneficial effects on healing than shear movements. In addition, for the latter, translational shearing of a fracture gap will more likely result in healing delays than torsional shearing for comparable loading magnitudes [17]. Furthermore, the optimal mechanical stability for osteosynthesis devices was calculated by interpolating between many different results from healing experiments in sheep.

For 96 combinations of axial and shear stiffness values, a numerical simulation of the predicted bending stiffness of healed sheep bones was calculated [17] and compared with the outcome from 14 sheep studies (Table 6.1). This allowed the creation of a map that indicates under which stiffness values a good healing outcome can be expected (Fig. 6.5). Table 6.1 lists the data of the 14 experimental studies shown in Fig. 6.5, Table 6.2 lists the conditions for exemplary simulations for

Table 6.1 Literature data of sheep tibia fracture healing experiments, outcome and stiffness of fixation

Author	Fixation device	Healing outcome (% control bone)	Axial stiffness (N/mm)	Shear stiffness (N/mm)
1. Epari et al. [18]	External fixator	61.5	2540	164
2. Epari et al. [18]	External fixator	83	2177	433
3. Epari et al. [18]	External fixator	68.2	1523	374
4. Epari et al. [18]	External fixator	66.3	1479	344
5. Epari et al. [18]	Unreamed nail	52.8	1213	139
6. Epari et al. [18]	Angle-sta-ble nail	64.1	2762	469
7. Bottlang et al. [19]	Locked plate	42	3922[a]	2500[a]
8. Bottlang et al. [19]	Far cortical Locked plate	67	628[b]	600[b]
9. Schell et al. [20]	External fixator	14	650	50[c]
10. Wolf et al. [21]	External fixator	60	183	170
11. Hente et al. [22]	External fixator	24	1666	220[c]
12. Bishop et al. [23]	External fixator	60–69	498	220[c]
13. Goodship [24]	External fixator	Bony bridging	500	350[c]
14. Goodship [24]	External fixator	Bony bridging	700	350[c]

Healing periods 9 weeks, except nr. 10 and 11 (6 weeks), nr.12 (8 weeks) and nr.13 and 14 (12 weeks)

Table adapted and reprinted by permission from [17]

[a]Shear and axial stiffness numerically calculated (adapted and reprented by permission by PlosOne) [17]

[b]Axial stiffness prior to bony contact of the pins, shear stiffness estimated with 20% error

[c]Shear stiffness assumed based on comparable devices [17] 20% error estimated

selected stiffness combinations and Table 6.3 describes the outcome quality of the various colored areas in the calculated map.

The numerical fracture healing simulation can deliver important information about the mechanical influences on the healing processes and can be used to create optimal fixations and investigate other influences, e.g. biological or genetic factors that influence the healing progress in animal studies. Furthermore, these results can

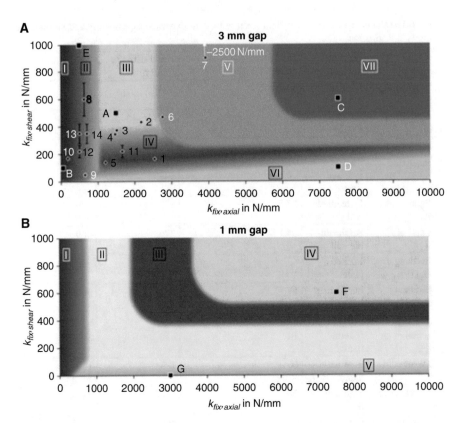

Fig. 6.5 Qualitative characteristic maps of healing outcome depending on the fixation stiffness in the axial (kfix,axial) and shear (kfix, shear) directions for (**a**) 3 mm fracture gap; (**b**) 1 mm fracture gap. Numbers refer to the animal experiments listed in Table 6.1. Letters indicate positions of the exemplary simulation results in Table 6.2. Roman numerals refer to areas of different healing outcomes as explained in Table 6.3 (reprinted by permission from PlosOne) [17]

Table 6.2 Exemplary simulations for 3 and 1 mm fracture gap sizes (Fig. 6.5)

Case	Conditions	Axial stiffness (N/mm)	Shear stiffness (N/mm)	Gap size (mm)
A	Optimal	1500	500	3
B	Overly flexible	50	100	3
C	Overly rigid	7500	600	3
D	Shear	7500	100	3
E	Axial	500	1000	3
F	Advantageous	7500	600	1
G	Disadvantageous	3500	1	1

Table 6.3 Comments to the qualitative characteristic maps in Figs. 6.5 and 6.6 (adapted and reprinted by permission by PlosOne)

	3 mm gap	1 mm gap
I	Non-union	Delayed healing
II	Slightly delayed healing	Slightly delayed healing
III	Optimal healing	Rapid healing
IV	Slightly delayed healing	Optimal healing
V	Suboptimal healing	Delayed healing
VI	Delayed healing	
VII	Unfavorable healing	

For further information Steiner et al. [17]

also serve as an orientation for the clinical treatment of fractures as well as for the development of new osteosynthesis devices

6.3   Conclusion for the Improvement of Clinical Fracture Healing

Because the sheep bone (i.e. tibia) size, shape, loading and bone healing time are of the same order of magnitude as those in humans, one can assume that the optimal fixation stiffness is similar in sheep and humans [18, 25].

Therefore, it is assumed that one can approximately use the stiffness map generated for sheep to predict patient healing outcome with certain fixation stiffness characteristics.

The optimal conditions for rapid callus healing of a diaphyseal fracture with a 3 mm fracture gap is fixation with a shear stiffness of >300 N/mm and an axial stiffness of 1000–2500 N/mm (Fig. 6.6) [3]. Deviations from this stiffness combination lead to delayed bone healing [17]. Using a different analytical approach, Epari et al. [18] found similar results, with an optimal shear stiffness of >380 N/mm and axial stiffness range of 1600–2400 N/mm [3].

These conditions are best meat by reamed nails with high axial [26] (1758–2282 N/mm) and shear stiffness.

Although the axial stiffness found in unreamed nails (9–12 mm diameter) for patients was acceptable (723–2300 N/mm) [26, 27], the shear stiffness was detrimental (131–224 N/mm). For external fixators, the axial stiffness was very low (148–510 N/mm) [28–32] whereas the shear stiffness was of a similar order than the axial stiffness. The axial stiffness of locking plates varies tremendously depending on the plate to bone surface distance and the number of screws used (85–2900 N/mm) [19, 33]. Extremely high axial stiffness of plates with opposite bone support (10,000 N/mm) can suppress bone formation because of the lack of strain stimulus underneath the plate [34]. Stiffer fixations for smaller fracture gaps (i.e. 1 mm) are advantageous because of increasing tissue strain with decreasing gap size for the same stiffness [3, 35].

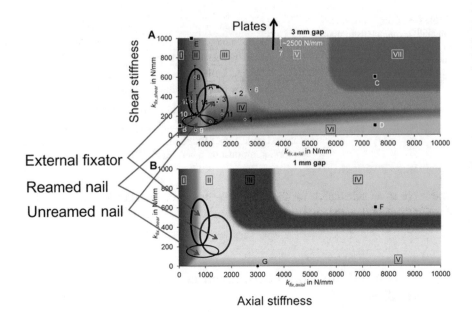

Fig. 6.6 Figure 6.5 extended with ellipses indicating the range of axial and shear stiffness published in clinical fracture fixation studies with external fixators and reamed and unreamed nails. Shear stiffness of plates is outside of the range shown in the map. (**a**) 3 mm fracture gap; (**b**) 1 mm fracture gap [17]. (Adapted and reprinted by permission from Elsevier). For an explanation of the numbers and letters, see Fig. 6.5 [3]

In contrast to the above-described method using stiffness data from a large number of fracture healing studies to predict the healing outcome for certain fixation cases, numerical simulation of individual fracture fixation cases can be applied to determine the optimal biomechanical conditions for fracture fixations.

Wehner et al. [16] applied numerical simulations to a clinical fracture fixation case, simulating the influence on the fracture healing time course of the stiffness of a human tibia fracture fixation. Prediction of the healing time by the numerical simulation was strongly influenced by the characteristics of the external fixator design parameters. The most significant effect on the healing time was displayed by the free bending length of the fixator screws. Reducing this from 40 to 20 mm reduced the predicted healing time by 50% [3, 16].

Wehner et al. [36] in a further study simulated the effect of the intramedullary nail diameter and material on human tibia healing time for different fracture gaps. They reported that to achieve the shortest healing time (assuming similar vascular conditions for all unreamed nails), the diameter of the unreamed intramedullary nails should be as large as possible and be made of stiffer stainless steel rather than flexible titanium. When stabilizing a tibia fracture using a 9 mm diameter titanium nail, an approximately 40-week healing time was calculated, whereas an 11 mm stainless steel nail displayed a shorter, approximately 15-week healing time. The healing time was considerably longer when increasing the gap size from 3 to 5 mm, particularly for smaller nail diameters [3].

Although clinically relevant results were produced using these numerical simulations of individual cases, which can be applied to optimize implant stiffness, they are extremely difficult to perform in clinical routine treatment for individual patients with various fracture situations and fixation systems.

To calculate the complex strain distribution in the healing zone and thus predict the healing outcome, such simulations require high-resolution 3D images of the fracture-healing zone and information about the individual fixation stiffness as well as loading of the patient [3, 37].

In contrast to the bone healing studies with healthy young animals, standardized fractur models and numerical simulations based on optimal healing rules and boundary conditions, the bone healing in patients is frequently impaired by other circumstances that can interact with the biomechanical situation.

Some of these circumstances are well investigated, including the degree of open fractures and fracture gap size, whereas little is known about others like individual healing capacity, age or infection.

The first animal studies demonstrated that there is an interaction between age and mechanical stability of fracture fixation leading to different results depending on age and stiffness of fixation. Whereas the strength of the healed bones in a young rat group was greater in the rigidly fixed group than in the semi-rigidly fixed group the old rats exhibited the reverse outcome [38]. An analysis of more than a hundred genes up-regulated or down-regulated in the fracture hematoma exhibited differential expression and interaction between age and fixation stability [39]. This indicates the highly complex interaction between genes and molecules and the biomechanical environment as reported in other studies [40–42].

Although the experimental studies regarding the effect of infections show some limitations, their results indicate that a more stable fracture fixation is a very important factor in the prevention and treatment of infected fractures [43, 44].

The above-described general mechano-biological knowledge appears to permit some conclusions for the optimization of human fracture fixation. When it becomes possible to transfer the recommendations concluded from the mechanobiological research to the clinical operative treatment of fractures, most of the still reported delayed healing could be avoided.

There might be however still some fractures that are successfully stabilized but display an impaired healing. For these cases, an analysis of possible reasons for delay of healing and an alternative or additional treatment might be applied. A number of physical (Chap. 5) and biological treatments are investigated [45, 46]. None of them demonstrate a high level of evidence for a significant treatment success and were broadly used in routine fracture treatment to date. However, when new future therapies will be successfully used, it will be particularly important to stabilize the fractured bones biomechanically optimal. Otherwise, these treatments will be constrained in their effectiveness.

This book attempted to describe the basic science of mechanobiology of the fracture healing process and the most important biomechanical factors effecting the healing outcome of fracture treatment. The application of this knowledge should enable the surgeon to perform the optimal treatment.

References

1. Pauwels, F. 1960. A new theory on the influence of mechanical stimuli on the differentiation of supporting tissue. The tenth contribution to the functional anatomy and causal morphology of the supporting structure. *Zeitschrift für Anatomie und Entwicklungsgeschichte* 121: 478–515.
2. Perren, S.M. 1979. Physical and biological aspects of fracture-healing with special reference to internal-fixation. *Clinical Orthopaedics and Related Research* 138: 175–196.
3. Claes, L. 2021. Improvement of clinical fracture healing - what can be learned from mechano-biological research? *Journal of Biomechanics* 115: 110148.
4. DiGioia, A.M.I., E.J. Cheal, and W.C. Hayes. 1986. Three-dimensional strain fields in a uniform osteotomy gap. *Journal of Biomechanical Engineering* 108: 273–280.
5. Carter, D.R. 1987. Mechanical loading history and skeletal biology. *Journal of Biomechanics* 20 (11-12): 1095–1109.
6. Carter, D.R., P.R. Blenman, and G.S. Beaupre. 1988. Correlations between mechanical stress history and tissue differentiation in initial fracture healing. *Journal of Orthopaedic Research* 6 (5): 736–748.
7. Carter, D.R., et al. 1998. Mechanobiology of skeletal regeneration. *Clinical Orthopaedics and Related Research* 355: 41–55.
8. Claes, L.E., and C.A. Heigele. 1999. Magnitudes of local stress and strain along bony surfaces predict the course and type of fracture healing. *Journal of Biomechanics* 32 (3): 255–266.
9. Claes, L.E., et al. 1998. Effects of mechanical factors on the fracture healing process. *Clinical Orthopaedics and Related Research* 355: 132–147.
10. Prendergast, P.J., R. Huiskes, K. Soballe, and E.S.B. Research Award. 1996. Biophysical stimuli on cells during tissue differentiation at implant interfaces. *Journal of Biomechanics* 30 (6): 539–548.
11. Claes, L. 2017. Mechanobiology of fracture healing. *Unfallchirurg* 120 (1): 13.
12. Ament, C., and E.P. Hofer. 2000. A fuzzy logic model of fracture healing. *Journal of Biomechanics* 33: 961–968.
13. Garcia-Aznar, J.M., et al. 2007. Computational simulation of fracture healing: influence of interfragmentary movement on the callus growth. *Journal of Biomechanics* 40 (7): 1467–1476.
14. Lacroix, D., and P.J. Prendergast. 2002. A mechano-regulation model for tissue differentiation during fracture healing: analysis of gap size and loading. *Journal of Biomechanics* 35 (9): 1163–1171.
15. Simon, U., et al. 2011. A numerical model of the fracture healing process that describes tissue development and revascularisation. *Computer Methods in Biomechanics and Biomedical Engineering* 41 (1): 79–93.
16. Wehner, T., et al. 2010. Influence of the fixation stability on the healing time - a numerical study of a patient-specific fracture healing process. *Clinical biomechanics* 25 (6): 606–612.
17. Steiner, M., et al. 2014. Numerical simulation of callus healing for optimization of fracture fixation stiffness. *PLoS One* 9 (7): e101370.
18. Epari, D.R., et al. 2007. Timely fracture-healing requires optimization of axial fixation stability. *The Journal of Bone and Joint Surgery. American Volume* 89 (7): 1575–1585.
19. Bottlang, M., et al. 2009. Far cortical locking can reduce stiffness of locked plating constructs while retaining construct strength. *The Journal of Bone and Joint Surgery. American Volume* 91 (8): 1985–1994.

20. Schell, H., et al. 2008. Mechanical induction of critically delayed bone healing in sheep: Radiological and biomechanical results. *Journal of Biomechanics* 41 (14): 3066–3072.
21. Wolf, S., et al. 1998. The effects of external mechanical stimulation on the healing of diaphyseal osteotomies fixed by flexible external fixation. *Clinical Biomechanics* 13 (5): 359–364.
22. Hente, R.F.B., et al. 2004. The influence of the cyclic compression and distraction on the healing of experimental tibial fractures. *Journal of Orthopaedic Research* 22 (4): 709–715.
23. Bishop, N.E., et al. 2006. Shear does not necessarely inhibit bone healing. *Clinical Orthopaedics and Related Research* 443: 307–314.
24. Goodship, A.E. 1993. The role of fixator frame stiffness in the control of fracture healing. An experimental study. *Journal of Biomechanics* 26 (9): 1027–1035.
25. Wissing, H., K.M. Stürmer, and G. Breidenstein. 1990. Die Wertigkeit verschiedener Versuchstierspezies für experimentelle Untersuchungen am Knochen. *Hefte zur Unfallheilkunde* 212: 479–488.
26. Schandelmaier, P., C. Krettek, and H. Tscherne. 1996. Biomechanical study of nine different tibia locking nails. *Journal of Orthopaedic Trauma* 10 (1): 37–44.
27. Penzkofer, R., et al. 2009. Influence of intramedullary nail diameter and locking mode on the stability of tibial shaft fracture fixation. *Archives of Orthopaedic and Trauma Surgery* 129 (4): 525–531.
28. Gardner, T.N., et al. 1994. Three-dimensional movement at externally fixated tibial fractures and osteotomies during normal patient function. *Clinical biomechanics* 9 (1): 51–59.
29. Höntzsch, D., et al. 1993. Die begleitende Fibulaosteosynthese bei der kompletten Unterschenkelfraktur. *Dental Traumatology* 9: 110.
30. Cunningham, J.L., M. Evans, and J. Kenwright. 1989. Measurement of fracture movement in patients treated with unilateral external skeletal fixation. *Journal of Biomedical Engineering* 11 (2): 118–122.
31. Gasser, B., et al. 1990. Stiffness characteristics of the circular Ilizarov device as opposed to conventional external fixators. *Journal of Biomechanical Engineering* 112 (1): 15–21.
32. Claes, L. 1991. Measuring bone healing in osteosynthesis with external fixator using the Fraktometer FM 100. *Chirurg* 62 (4): 354–355.
33. Stoffel, K., et al. 2003. Biomechanical testing of the LCP--how can stability in locked internal fixators be controlled? *Injury* 34 (2): 11–19.
34. Röderer, G., et al. 2014. Delayed bone healing following high tibial osteotomy related to increased implant stiffness in locked plating. *Injury* 45 (10): 1648–1652.
35. Perren, S.M., and J. Cordey. 1977. Tissue differences in fracture healing (author's transl). *Unfallheilkunde* 80 (5): 161–164.
36. Wehner, T., et al. 2012. Optimization of intramedullary nailing by numerical simulation of fracture healing. *Journal of Orthopaedic Research* 30 (4): 569–573.
37. Shefelbine, S.J., et al. 2005. Prediction of fracture callus mechanical properties using micro-CT images and voxel-based finite element analysis. *Bone* 36 (3): 480–488.
38. Strube, P., et al. 2008. Influence of age and mechanical stability on bone defect healing: age reverses mechanical effects. *Bone* 42 (4): 758–764.
39. Ode, A., et al. 2014. Interaction of age and mechanical stability on bone defect healing: an early transcriptional analysis of fracture hematoma in rat. *PLoS One* 9 (9): e106462.
40. Le, A.X., et al. 2001. Molecular aspects of healing in stabilized and non-stabilized fractures. *Journal of Orthopaedic Research* 19 (1): 78–84.
41. Lienau, J., et al. 2010. Insight into the molecular pathophysiology of delayed bone healing in a sheep model. *Tissue Engineering. Part A* 16 (1): 191–199.
42. Palomares, K.T., et al. 2009. Mechanical stimulation alters tissue differentiation and molecular expression during bone healing. *Journal of Orthopaedic Research* 27 (9): 1123–1132.

43. Foster, A.L., et al. 2021. The influence of biomechanical stability on bone healing and fracture-related infection: the legacy of Stephan Perren. *Injury* 52 (1): 43–52.
44. Worlock, P., et al. 1994. The prevention of infection in open fractures: an experimental study of the effect of fracture stability. *Injury* 25 (1): 31–38.
45. Einhorn, T.A. 1995. Enhancement of fracture-healing. *Journal of Bone and Joint Surgery* 77 (6): 940–956.
46. Schlickewei, C.W., et al. 2019. Current and future concepts for the treatment of impaired fracture healing. *International Journal of Molecular Sciences* 20: 22.

Correction to: Basic Biomechanical Factors Affecting Fracture Healing

Correction to:
Chapter 4 in: L. E. Claes, *Mechanobiology of Fracture Healing*, SpringerBriefs in Bioengineering, https://doi.org/10.1007/978-3-030-94082-9_4

The original version of this chapter was inadvertently published with incorrect Figure 4.6c, which has been corrected.

The correct Fig. 4.6 is presented below

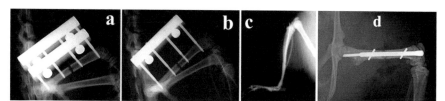

Fig. 4.6 Frequently used fixation techniques for rat fractures. For the femur most often an unilateral external fixator is used (**a** stiff fixator, **b** flexible fixator) [43, 62] (reprinted by permission from John Wiley and Sons). For the tibia frequently injection needles or steel wires are used (**c**) as an intamedullary nail (reprinted by permission from John Wiley and Sons [63]. For the femur interlocking nails are now available (**d**) (reprinted by permission from John Wiley and Sons) [64]

The updated version of this chapter can be found at
https://doi.org/10.1007/978-3-030-94082-9_4

Printed in the United States
by Baker & Taylor Publisher Services